AF701903

INTRODUCTION TO STATISTICAL FIELD THEORY

Knowledge of the renormalization group and field theory is a key part of physics, and is essential in condensed matter and particle physics. Written for advanced undergraduate and beginning graduate students, this textbook provides a concise introduction to this subject.

The textbook deals directly with the loop expansion of the free energy, also known as the background field method. This is a powerful method, especially when dealing with symmetries and statistical mechanics. In focusing on free energy, the author avoids long developments on field theory techniques. The necessity of renormalization then follows.

EDOUARD BRÉZIN is Honorary Professor of Physics at the Ecole Polytechnique, and Emeritus Professor at the Ecole Normale Supérieure (Paris). A former President of the French Academy of Sciences, he is also a Foreign Member of the National Academy of Sciences (USA), the American Academy of Arts and Sciences, the Royal Society, and Academia Europaea.

INTRODUCTION TO STATISTICAL FIELD THEORY

EDOUARD BRÉZIN
Ecole Normale Supérieure, Paris

CAMBRIDGE
UNIVERSITY PRESS

CAMBRIDGE
UNIVERSITY PRESS

Shaftesbury Road, Cambridge CB2 8EA, United Kingdom

One Liberty Plaza, 20th Floor, New York, NY 10006, USA

477 Williamstown Road, Port Melbourne, VIC 3207, Australia

314–321, 3rd Floor, Plot 3, Splendor Forum, Jasola District Centre, New Delhi – 110025, India

103 Penang Road, #05–06/07, Visioncrest Commercial, Singapore 238467

Cambridge University Press is part of Cambridge University Press & Assessment, a department of the University of Cambridge.

We share the University's mission to contribute to society through the pursuit of education, learning and research at the highest international levels of excellence.

www.cambridge.org
Information on this title: www.cambridge.org/9780521193030

© E. Brézin 2010

This publication is in copyright. Subject to statutory exception and to the provisions of relevant collective licensing agreements, no reproduction of any part may take place without the written permission of Cambridge University Press & Assessment.

First published 2010

A catalogue record for this publication is available from the British Library

ISBN 978-0-521-19303-0 Hardback

Cambridge University Press & Assessment has no responsibility for the persistence or accuracy of URLs for external or third-party internet websites referred to in this publication and does not guarantee that any content on such websites is, or will remain, accurate or appropriate.

Contents

Preface

These lecture notes do not attempt to cover the subject in its full extent. There are several excellent books that go much deeper into renormalization theory, or into the physical applications to critical phenomena and related topics. In writing these notes I did not mean either to cover the more recent and exciting aspects of the subject, such as quantum criticality, two-dimensional conformal invariance, disordered systems, condensed matter applications of the AdS/CFT duality borrowed from string theory, and so on.

A knowledge of the renormalization group and of field theory remains a necessary part of today's physics education. These notes are simply an introduction to the subject. They are based on actual lectures, which I gave at Sun Yat-sen University in Guangzhou in the fall of 2008. In order not to scare the students, I felt that a short text was a better introduction. There are even several parts that can be dropped by a hasty reader, such as GKS inequalities or high-temperature series. However, high-T series lead to an easy way of connecting geometrical criticality, such as self-avoiding walks and polymers or percolation to physics. I have chosen not to use Feynman diagrams; not that I think that they are unnecessary, I have used them for ever. But since I did not want to require a prior exposition to quantum field theory, I would have had to deal with a long detour, going through connected diagrams, one-particle irreducibility, and so on. I have chosen instead to base everything on the loop expansion of the free energy, not going here beyond one loop. This method, known nowadays as the background field method, is powerful (specially when dealing with symmetries) and natural from the viewpoint of statistical mechanics. (However, the more technical Chapter 13, on the renormalization of the non-linear sigma model, is aimed at readers who have some familiarity with diagrams.)

In spite of the briefness of these notes I wanted to make it clear why, after K. Wilson's work, not only were critical phenomena understood but the understanding of the meaning of renormalizability in quantum field theory changed drastically. It became clear that a beautiful renormalizable theory, such as quantum electrodynamics, was merely an *effective* theory, rather than a theory able to describe electromagnetism from astronomical distances down to vanishingly small length scales. This does not deprive QED from its exceptional beauty and its astounding agreement with experiment. In the post-Wilson analysis its renormalizability results from the fact that, like critical phenomena, the present day experiments, even at the highest presently available accelerator energies, deal with very large length scales in comparison to those at which new physics must occur. Why 'must'? It is because, unlike QCD, QED lacks 'asymptotic freedom', with the consequence that QED is 'trivial', meaning that it is only for a vanishingly small charge of the electron that it could deal with the smallest length scales. So viewing such theories, in the light of critical phenomena, told us that there has to be new physics at short distance.

Many books overlap part or most of the material of these lectures; among a long list, here is a short selection:

- John Cardy, *Scaling and Renormalization in Statistical Physics* (Cambridge: Cambridge University Press, 1996).
- Giorgio Parisi, *Statistical Field Theory* (New York: Addison-Wesley, 1988).
- J. Zinn-Justin, *Quantum Field Theory and Critical Phenomena*, 3rd edn (Oxford: Oxford University Press, 2002).
- J. Zinn-Justin, *Phase Transitions and the Renormalization Group* (Oxford: Oxford University Press, 2007).
- D. J. Amit and V. Martin-Mayor, *Field Theory, the Renormalization Group and Critical Phenomena* (Singapore: World Scientific, 2005).
- C. Itzykson and J. M. Drouffe, *Statistical Field Theory*, vols 1 and 2. (Cambridge: Cambridge University Press, 1989).

Note also the historical article based on K. Wilson's lectures in Princeton (1971–72): K. G. Wilson and J. Kogut, The renormalization group, *Phys. Rep.*, **12c** (1974) 75. Several books in the long series devoted to *Phase Transitions and Critical Phenomena*, edited by C. Domb and M. Green, and later by C. Domb and J. Lebowitz, will provide additional light on some of the topics of these lectures; see, e.g., vol. 6 of the series.

1

A few well-known basic results

This chapter is just a reminder of some basic results concerning equilibrium statistical mechanics and of a few algebraic techniques used in this book.

1.1 The Boltzmann law

For a system at equilibrium in contact with a heat bath (or thermostat) at temperature T, the configurations of the particles and the total energy are random variables. The equilibrium probability distribution for N identical particles confined in a box of volume V, whose dynamics are governed by a Hamiltonian H, is given by the Boltzmann Gibbs distribution

$$\rho = \frac{1}{Z} \mathrm{e}^{-\beta H}, \tag{1.1}$$

in which β is related to the temperature by

$$\beta = \frac{1}{kT}. \tag{1.2}$$

1.1.1 The classical canonical ensemble

For classical particles, in three dimensions, ρ is a probability measure in the $6N$-dimensional phase space (p_a, q_a), $a = 1 \ldots 3N$ and the expectation value of an observable $A(p, q)$ is given by

$$\langle A \rangle = \int \mathrm{d}\tau A(p, q) \rho(p, q), \tag{1.3}$$

in which $\mathrm{d}\tau$ is the measure $\mathrm{d}\tau = \frac{1}{h^{3N}} \frac{1}{N!} \prod_1^{3N} \mathrm{d}p_a \mathrm{d}q_a$. The integrals over the positions q_a are such that every particle is confined in a box of volume V.

The factor $1/h^{3N}$, in which h has the dimension of an action (i.e., ML^2T^{-1}), makes $\mathrm{d}\tau$ dimensionless. Any constant with that dimension would work but the classical limit of quantum statistical mechanics provides Planck's constant, $h = 2\pi\hbar$.

The factor $1/N!$ is also of quantum origin: Pauli's principle allows only for one-dimensional representations of the permutation group of N particles, completely symmetric (bosons) or completely antisymmetric (fermions). This selects only one state out of the degenerate $N!$ states obtained by permutations of one of them.

The normalization is fixed by $\langle 1 \rangle = 1$, which gives the partition function Z:

$$Z(\beta, N, V) = \int \mathrm{d}\tau \mathrm{e}^{-\beta H}. \tag{1.4}$$

1.1.2 The quantum canonical ensemble

The density matrix ρ, given by (1.1), is an operator in the Hilbert space of symmetric states for integer spin particles, or antisymmetric states for half-integer spins, for N particles confined in a box of volume V. The expectation value of an observable A is given by

$$\langle A \rangle = \mathrm{Tr}(\rho A) = \frac{1}{Z}\mathrm{Tr} A \mathrm{e}^{-\beta H} \tag{1.5}$$

and thus the partition function is given by

$$Z(\beta, N, V) = \mathrm{Tr}\, \mathrm{e}^{-\beta H}. \tag{1.6}$$

If the eigenvalues of the N-body Hamiltonian are labelled as E_i, then

$$Z = \sum_i \mathrm{e}^{-\beta E_i}. \tag{1.7}$$

If the energy E_i has a degeneracy w_i then

$$Z = \sum{}' \mathrm{e}^{-\beta(E_i - T S_i)}, \tag{1.8}$$

in which $S_i = k \log w_i$ and the last sum runs over distinct energies. This expression shows that the dominant contributions are those that minimize the combination $E - TS$, a competition between energy and entropy to which we shall return in the next section.

Exercise 1

Quantum effects arise when the typical de Broglie wavelength associated with a particle becomes comparable to the interparticle distance. Estimate the temperature below which quantum effects should be taken into account for a gas of nitrogen of atmospheric density.

1.1.3 The grand canonical ensemble

If the system, in contact with a heat bath, can also exchange particles with a reservoir at temperature T and chemical potential μ, the number of particles is also a random variable. In the simple case in which the Hamiltonian H_N does not change the number of particles, the probability distribution is given by a collection of ρ_N given by

$$\rho_N = \frac{1}{Z_G} \mathrm{e}^{\alpha N - \beta H_N}, \tag{1.9}$$

with

$$\mu = \frac{\alpha}{\beta}, \tag{1.10}$$

normalized by

$$Z_G(\alpha, \beta, V) = \sum_N \mathrm{e}^{\alpha N} \mathrm{Tr}\, \mathrm{e}^{-\beta H_N},$$

in which V is the volume of the box in which the particles are confined. (If the Hamiltonian does not conserve the number of particles, it is necessary to use the Fock space; this will not be needed within these lectures.)

1.2 Thermodynamics from statistical physics

The canonical free energy is given by

$$F(\beta, N, V) = -\frac{1}{\beta} \log Z. \tag{1.11}$$

Exercise 2

Show that the pressure, the entropy and the chemical potential of the system can all be related to the partition function. Compute the partition function for a classical gas of non-interacting particles.

1.2.1 The thermodynamic limit

The *thermodynamic limit* is the limit in which N and V go to infinity with a fixed ratio $\nu = N/V$. In this limit one can show that, for particles with short-range interactions, the canonical $\log Z$, and thus F are extensive, namely

$$\lim_{N\to\infty, V\to\infty} \bigg|_{N/V=\nu} \left(\frac{1}{N} \log Z \right) \tag{1.12}$$

exists and is a function of the two intensive variables ν and β. Similarly, for the grand canonical ensemble, $\lim_{V\to\infty} 1/V \log Z_G$ exists and is a function of the intensive variables, temperature and chemical potential.

Exercise 3

Verify this extensivity for N free classical particles in a box. Reminder: Stirling's formula $N! = \sqrt{2\pi N}\left(\frac{N}{e}\right)^N \left(1 + O\left(\frac{1}{N}\right)\right)$.

For charged particles, such as electrons with Coulomb interactions, the thermodynamic limit exists, provided that (a) the system is neutral, i.e., the charge of the ions compensates the charge of the electrons, (b) the system is quantum mechanical, (c) Pauli's principle is taken into account.[1]

Exercise 4

Assume that the potential energy of N interacting classical particles is a homogeneous function

$$V(\lambda q_1 \cdots \lambda q_{3N}) = |\lambda|^s V(q_1 \cdots q_{3N}).$$

Show that the pressure $p(v, T)$*, where* $v = N/V$*, satisfies the relation*

$$p(v, T) = T^{1-3/s}\varphi(vT^{3/s}).$$

Assume that at low temperature T_0 *the isotherm in the* (p, V) *plane presents a phase transition between two phases of different densities. Can there be a critical point for this phase transition, i.e., a temperature at which the transition between the two phases disappears?*

1.3 Gaussian integrals and Wick's theorem

1. *One variable*

$$\int_{-\infty}^{+\infty} \mathrm{d}x \mathrm{e}^{-\frac{1}{2}ax^2} = \sqrt{\frac{2\pi}{a}}. \tag{1.13}$$

2. *n variables*

$$\int_{R^n} \mathrm{d}x_1 \cdots \mathrm{d}x_n \mathrm{e}^{-\frac{1}{2}\sum x_i A_{ij} x_j} = \frac{(2\pi)^{n/2}}{\sqrt{\det A}}. \tag{1.14}$$

$A = A^t$ is here a real symmetric matrix with positive eigenvalues. It can thus be diagonalized by an orthogonal transformation ω, i.e., $A = \omega^t D\omega$, in which D is the diagonal matrix of the eigenvalues $(a_1, \ldots, a_n)$ of A. The change of variables $\omega x = y$ whose Jacobian $(|\det \omega|^{-1})$ is equal to one, leads to the solution.

3. *n variables in a source*

$$\frac{\int_{R^n} \mathrm{d}x_1 \cdots \mathrm{d}x_n \mathrm{e}^{-\frac{1}{2}\sum x_i A_{ij} x_j + \sum b_i x_i}}{\int_{R^n} \mathrm{d}x_1 \cdots \mathrm{d}x_n \mathrm{e}^{-\frac{1}{2}\sum x_i A_{ij} x_j}} = \mathrm{e}^{\frac{1}{2}\sum b_i A^{-1}_{ij} b_j}. \tag{1.15}$$

Translate $x = y + A^{-1}b$.

[1] J. Lebowitz and E. Lieb, *Phys. Rev. Lett.*, **22** (1969) 631.

4. *Wick's theorem*
Apply to (1.15) the operation $\frac{\partial}{\partial b_{i_1}} \cdots \frac{\partial}{\partial b_{i_{2n}}}$ and then set all the $b_i = 0$. The l.h.s. gives

$$\langle x_{i_1} \cdots x_{i_{2n}} \rangle = \frac{\int_{R^n} dx_1 \cdots dx_n e^{-\frac{1}{2}\sum x_i A_{ij} x_j} x_{i_1} \cdots x_{i_{2n}}}{\int_{R^n} dx_1 \cdots dx_n e^{-\frac{1}{2}\sum x_i A_{ij} x_j}}. \tag{1.16}$$

Applying this to the r.h.s. of (1.15) we can limit ourselves to the term $\frac{1}{(n)!2^n} \times \left(\sum b_i A^{-1}_{ij} b_j\right)^n$; indeed terms of lower degree in the expansion of the exponential will give zero by differentiation; terms of higher degree will give zero because they are left with b and vanish at $b = 0$. Therefore,

$$\langle x_{i_1} \cdots x_{i_{2n}} \rangle = \frac{\partial}{\partial b_{i_1}} \cdots \frac{\partial}{\partial b_{i_{2n}}} \frac{1}{(n)!2^n} \left(\sum b_i A^{-1}_{ij} b_j\right)^n. \tag{1.17}$$

Define a complete pairing of the $\frac{\partial}{\partial b}$ such that each $\frac{\partial}{\partial b_i}$ has a partner. For this particular pairing, the two paired differentiations go to the same $\sum$, but there are $n!$ ways of associating the sums and the chosen pairing. Once this association is made, one has simply to note that

$$\frac{\partial}{\partial b_k}\frac{\partial}{\partial b_l} \sum b_i A^{-1}_{ij} b_j = 2A^{-1}_{kl}.$$

Therefore the $n!2^n$ cancels and we are left with the result, known as Wick's theorem, for Gaussian integrals.

$$\langle x_{i_1} \cdots x_{i_{2n}} \rangle = \sum_{\text{pairings}} \prod_{\substack{\text{each}\\ \text{pair}}} A^{-1}_{i_a i_b}. \tag{1.18}$$

Exercise

Compute the integral

$$I = \frac{\int_{R^2} dxdy\, x^4 y^2 e^{-(x^2+xy+2y^2)}}{\int_{R^2} dxdy e^{-(x^2+xy+2y^2)}}.$$

Answer

$$I = \frac{144}{343},$$

since

$$I = 3(\langle xx \rangle)^2 \langle yy \rangle + 12 \langle xx \rangle (\langle xy \rangle)^2,$$

in which

$$\langle xx \rangle = A^{-1}_{11} \quad \langle xy \rangle = A^{-1}_{12} \quad \langle yy \rangle = A^{-1}_{22},$$

with

$$A = \begin{pmatrix} 2 & 1 \\ 1 & 4 \end{pmatrix} \quad A^{-1} = \frac{1}{7}\begin{pmatrix} 4 & -1 \\ -1 & 2 \end{pmatrix}.$$

1.4 Functional derivatives

A functional $F\{f\}$ is an application from a space of functions f to a complex or real number F. For instance the action integral for the motion of a particle, located at the position $q(t)$ at time t, with potential energy $V(q)$ is the functional of the trajectory given by

$$S\{q\} = \int_{t_1}^{t_2} \mathrm{d}t \left[\frac{m}{2}\dot{q}^2 - V(q)\right]. \tag{1.19}$$

Let us work with functions f of a single real variable x (the generalization to functions of more variables is immediate). The derivative of the functional with respect to $f(x)$ at $x = x_0$ is defined as follows. Let us consider an increment $\epsilon\delta_\eta(x - x_0)$; the function $\delta_\eta(x)$ is centred at the origin, and it has a width η; it is normalized to one, i.e., $\int_{\mathcal{R}} \delta_\eta(x)\mathrm{d}x = 1$. When η goes to this zero, this increment approaches the Dirac distribution $\delta(x)$. (For instance $\delta_\eta(x) = \frac{1}{\eta\sqrt{2\pi}}\mathrm{e}^{-x^2/2\eta^2}$.) One computes next the increment of the functional

$$\Delta F = F\{f + \epsilon\delta_\eta(x - x_0)\} - F\{f\}. \tag{1.20}$$

The functional derivative of F at x_0 is defined as

$$\left.\frac{\delta F}{\delta f}\right|_{x_0} = \lim_{\eta\to 0}\lim_{\epsilon\to 0}\frac{\Delta F}{\epsilon}. \tag{1.21}$$

The limits have to be taken in the order indicated: if we let ϵ go to zero first, we avoid non-linearities in δ_η. In the opposite order we would encounter powers of δ_η, which do not have a limit when η goes to zero.

Let us apply this to the above action functional:

$$\begin{aligned} &\lim_{\epsilon\to 0}\frac{1}{\epsilon}[S\{q(t) + \epsilon\delta_\eta(t - t_0)\} - S\{q\}] \\ &\qquad = \int_{t_1}^{t_2} \mathrm{d}t[m\dot{q}\,\dot{\delta}_\eta(t - t_0) - V'(q)\delta_\eta(t - t_0)]. \end{aligned} \tag{1.22}$$

After an integration by parts of the first term one ends up with

$$\frac{\delta S}{\delta q}(t_0) = -m\ddot{q}(t_0) - V'(q(t_0)) \tag{1.23}$$

and Newton's law is just given by the vanishing of this functional derivative for any t_0: the action is stationary (in fact a minimum) for the classical trajectory.

1.5 *d*-dimensional integrals

The rules are simple but they may surprise the reader who sees them for the first time. Whenever the dimension d is an integer, the d-dimensional integral is the ordinary integral over the whole space R^d. But, for arbitrary d, one applies the following rules:

(a) $\int \mathrm{d}^d q f(q+p) = \int \mathrm{d}^d q f(q)$,
(b) $\int \mathrm{d}^d q f(\lambda q) = |\lambda|^{-d} \int \mathrm{d}^d q f(q)$.

If q_1 is a d_1-dimensional vector and q_2 is a d_2-dimensional vector and $f(q) = g_1(q_1) g_2(q_2)$ with $d = d_1 + d_2$, then

(c) $\int \mathrm{d}^d q f(q) = \int \mathrm{d}^{d_1} q_1 g_1(q_1) \int \mathrm{d}^{d_2} q_2 g_2(q_2)$.

Consequences:

- From (b) the only finite solution to an integral, such as $\int \mathrm{d}^d q (q^2)^k$ is

$$\int \mathrm{d}^d q (q^2)^k = 0$$

  for any positive or negative real number k, including $k = 0$. Note that this integral never exists as an ordinary integral for integer dimensions. The consistency of this rule will be checked below.
- The same would apply to any scale-invariant integral, such as

$$\int \mathrm{d}^d q_1 \mathrm{d}^d q_2 (q_1^2)^k \left[(q_1 + q_2)^2\right]^l = 0.$$

- From (c)

$$\int \mathrm{d}^d q \; \mathrm{e}^{-q^2} = \left[\int_{-\infty}^{+\infty} \mathrm{d}x \mathrm{e}^{-x^2}\right]^d = \pi^{d/2}.$$

Let us use these rules to calculate simple integrals:

$$\begin{aligned}\int \mathrm{d}^d q (q^2+1)^{-k} &= \frac{1}{\Gamma(k)} \int \mathrm{d}^d q \int_0^\infty \mathrm{e}^{-\lambda(q^2+1)} \lambda^{k-1} \mathrm{d}\lambda \\ &= \frac{\pi^{d/2}}{\Gamma(k)} \int_0^\infty \mathrm{d}\lambda \lambda^{k-d/2-1} \mathrm{e}^{-\lambda} = \frac{\pi^{d/2}\Gamma(k-d/2)}{\Gamma(k)}.\end{aligned}$$

One can also compute this integral in 'spherical' coordinates:

$$\begin{aligned}\int \mathrm{d}^d q (q^2+1)^{-k} &= \frac{2\pi^{d/2}}{\Gamma(d/2)} \int_0^\infty \mathrm{d}x \; x^{d-1} \frac{1}{(x^2+1)^k} \\ &= \frac{2\pi^{d/2}}{\Gamma(d/2)} \frac{1}{2} \int_0^1 \mathrm{d}y \; y^{k-d/2-1} (1-y)^{d/2-1} \\ &= \frac{\pi^{d/2}\Gamma(k-d/2)}{\Gamma(k)}\end{aligned}$$

(change $1/(1+x^2) = y$). It is easy to verify on examples, such as $d=3$ and $k=2$, that whenever the integral exists in the ordinary sense it is indeed given by this result.

To check the consistency of rule (b) let us compute

$$J = \int \mathrm{d}^d q \frac{1}{q^2(q^2+1)}.$$

If we use spherical coordinates,

$$\begin{aligned} J &= \frac{2\pi^{d/2}}{\Gamma(d/2)} \int_0^\infty \mathrm{d}x\, x^{d-3} \frac{1}{x^2+1} = \frac{2\pi^{d/2}}{\Gamma(d/2)} \frac{1}{2} \int_0^1 \mathrm{d}y\, y^{1-d/2}(1-y)^{d/2-2} \\ &= \frac{\pi^{d/2}}{\Gamma(d/2)} \Gamma(2-d/2)\Gamma(d/2-1). \end{aligned}$$

It is easy to verify that, for $d = 3$, J exists as an ordinary integral and is indeed given by this result. Alternatively, using the identity $\frac{1}{q^2(q^2+1)} = \frac{1}{q^2} - \frac{1}{q^2+1}$, we find, from rule (b) and the above $k = 1$ result,

$$J = 0 - \pi^{d/2}\Gamma(1-d/2),$$

and it is easy to check that this coincides with the above result for J.

Additional references

Shang-Keng Ma, *Statistical Mechanics*, (Singapore: World Scientific, 1998).

L. P. Kadanoff, *Statistical Physics, Statics, Dynamics and Renormalization*, (Singapore: World Scientific, 2000).

On the existence of the thermodynamic limit:

David Ruelle, *Statistical Mechanics: Rigorous results*, (New York: Benjamin, 1969).

Kerson Huang, *Statistical Mechanics*, (New York: Wiley, 1963), Appendix C.

2

Introduction: order parameters, broken symmetries

2.1 Can statistical mechanics be used to describe phase transitions?

A phenomenological description of a phase transition does not raise any special difficulty a priori. For instance, to describe the solidification of a gas under pressure, one can make a simple theory for the gaseous phase, e.g., an ideal gas corrected by a few terms of the virial expansion. Then, for the solid, one can use the extraction energies of the atoms, and the vibration energies around equilibrium positions. These calculations will provide a thermodynamic potential for each phase. The line of coexistence between the two phases in the pressure–temperature plane will be determined by imposing the equality of the two chemical potentials $\mu_I(T, P) = \mu_{II}(T, P)$.

If this method may turn out to be useful in practice, it does not answer any of the questions that one can raise concerning the transition between the two states. Indeed the interactions between the molecules are not statistical in nature: they are independent of the temperature, or of the pressure; the Hamiltonian is a combination of kinetic energy and well-defined interaction potentials between pairs of molecules. How can one see in such a description, following the principles established by Boltzmann, Gibbs and their successors, that at equilibrium the same molecules can form a solid or a fluid, a superconductor, a ferromagnet, etc., without any modification of the interactions? It is so far from obvious that, for a long time, some believed that the principles of statistical mechanics had to be completed to allow for the possibility of a phase transition. It is only after Peierls' 1936 work (which will be reviewed below)[1] and Onsager's[2] solution of the two-dimensional Ising model, that it became manifest that the ordinary principles of statistical mechanics contained the possibility of phase transitions and

[1] R. Peierls, *Proc. Cambridge Phil. Soc.*, **32** (1936) 477.

[2] L. Onsager, Crystal statistics: a two-dimensional model with an order–disorder transition, *Phys. Rev.*, **65** (1944) 117.

critical points. (A critical point is the endpoint of a line of coexistence between two phases, such as the endpoint of the liquid–vapour line of coexistence in the (p, T) plane. Beyond this point there is only one phase, the fluid phase.)

These same contributions also demonstrated how singularities, such as discontinuities, or divergences of physical quantities at critical points, could arise in statistical mechanics, whereas one could have thought, a priori, that they were not possible. Indeed the thermodynamic properties are determined by the knowledge of the free energy, proportional to $\log Z$, with

$$Z = \sum_{C} \mathrm{e}^{-\beta E(C)}. \tag{2.1}$$

Each term of the sum is analytic in the temperature (except at zero temperature). Assuming that there is a finite number of distinct configurations, as will be the case in many situations that we are going to study, such as the Ising model, then the partition function is analytic in T, as well as the free energy $F = -kT \log Z$, unless Z vanishes. But Z does not vanish for real values of T, and thus F is non-singular on the real temperature axis. One can thus conclude that singularities cannot exist at any real non-zero temperature: in other words, phase transitions do not exist!

We shall see that the solution to this paradox is that the number N of constituents is so large that one almost always observes experimentally only the thermodynamic limit and the simplistic argument that I have just used fails in that limit.

2.2 The order–disorder competition

To make the discussion more concrete, I shall introduce immediately the Ising model, whose physical significance will be explained in the next section. We consider a periodic lattice with N sites in dimension d. To each site i one attaches a 'spin':

$$\sigma_i = \pm 1 \tag{2.2}$$

(for a true spin this would be simply the eigenvalues of one component of a spin $\frac{1}{2}$, up to a factor $\hbar/2$). Therefore, there are 2^N configurations $\mathcal{C}$

$$\mathcal{C} = (\sigma_1, \ldots, \sigma_N). \tag{2.3}$$

We now have to define the energy of a configuration $\mathcal{C}$. In the simplest model introduced by Ising, a student of Lenz, in 1925,[3] one assumes that only pairs of

[3] Ising solved the model in one dimension, in which there is no phase transition and conjectured (erroneously) on that basis that the model could not describe a phase transition.

neighbouring spins interact. In other words, given two spins σ_i and σ_j, their energy of interaction is

$$E(\sigma_i, \sigma_j) = \begin{pmatrix} 0 & \text{(if } i \text{ and } j \text{ are not nearest neighbours)} \\ -J & \text{(if } i \text{ and } j \text{ are nearest neighbours and } \sigma_i = \sigma_j) \\ +J & \text{(if } i \text{ and } j \text{ are nearest neighbours and } \sigma_i = -\sigma_j) \end{pmatrix}. \tag{2.4}$$

In other words, for two neighbouring sites i and j

$$E(\sigma_i, \sigma_j) = -J\sigma_i\sigma_j. \tag{2.5}$$

If J is positive (ferromagnetic interaction) the energy is lower for parallel spins. The energy for the whole lattice is the sum of the pair energies of neighbouring sites, i.e., for the configuration $\mathcal{C}$,

$$E(\mathcal{C}) = -J \sum_{\langle i,j \rangle} \sigma_i\sigma_j, \tag{2.6}$$

where, by convention, the bracket $\langle i, j \rangle$ limits the sum to all the pairs of nearest neighbours. The minimum energy (for $J > 0$) is reached for the two configurations in which all the spins are parallel,

$$E_{\min} = -JL, \tag{2.7}$$

in which L is the total number of links.

Exercise

Compute L/N for a hypercubic lattice in dimension d, N large.

Therefore, the state of the system at zero temperature will depend on the boundary conditions. To obtain a *pure state* one may 'freeze' the spins on the boundary. For instance, one may introduce a layer of external spins frozen to the value $\sigma = +1$ interacting with their neighbours. Then the configuration of minimum energy is $\sigma_i = +1$ for the whole lattice. There is, of course, a second pure state with $\sigma_i = -1$. Other boundary conditions, such as free boundary conditions, would lead, at zero temperature, to a *mixture* of the previous pure states.

At finite temperature, the sum over the 2^N configurations is

$$Z = \sum_{\mathcal{C}} \mathrm{e}^{-\beta E(\mathcal{C})}, \tag{2.8}$$

which can be written equivalently as

$$Z = \sum{}' w(E)\mathrm{e}^{-\beta E}, \tag{2.9}$$

in which the sum primed runs over distinct energies; $w(E)$ is the number of configurations of energy E and it is more convenient to use Boltzmann's entropy,

$$S(E) = k \log w(E). \tag{2.10}$$

Then

$$Z = \sum\nolimits' e^{-\beta(E - TS(E))}. \tag{2.11}$$

For large N the typical values of E and $S(E)$ are extensive, i.e., proportional to N. Therefore, the sum is peaked around the energy which minimizes $F(E) = E - TS(E)$, i.e., for energies such as

$$1 = T\frac{\partial S}{\partial E}.$$

One recovers the thermodynamic definition of temperature and the minimization of the free energy. This minimization reflects the competition between order and disorder.

At very low temperatures, the leading term in $F(E)$ is the energy, and the minimization of energy leads to an ordered state of parallel spins. At higher temperatures, one may get a lower free energy by allowing for higher values of the energy, provided the entropy is large to benefit from the negative term $-TS(E)$. The low-energy configurations have a small entropy. For instance, the degeneracy of the configuration obtained by turning one spin down, of energy $E_{\min} + 2dJ$ (hypercubic lattice in dimension d), is equal to N and the entropy per site $(1/N)k \log N$ vanishes for large N. However, at higher temperatures the leading configurations have a much higher energy (proportional to N) but this may be compensated by an even larger $TS(E)$, also proportional to N, and the disorder is then favoured.

This discussion is much too qualitative to be really conclusive. First of all, we have not yet defined the parameter that will allow us to characterize the state of order of the system. Then one could imagine a gradual disappearance of the ordered state when the temperature increases. Or else the zero-temperature ordered state could disappear immediately at any non-zero temperature. Therefore, we need to examine these questions in a more quantitative manner.

2.3 Order parameter, symmetry and broken symmetry

The energy $E(\mathcal{C})$ is invariant if all the spins are flipped simultaneously. One could then hastily conclude that the average spin at a given site i

$$\langle \sigma_i \rangle = \frac{1}{Z}\sum_{\mathcal{C}} \sigma_i \, e^{-\beta E(\mathcal{C})} \tag{2.12}$$

vanishes (from the interchange $\sigma \rightarrow -\sigma$ for all sites). However, we have to face the problem of boundary conditions that we have encountered earlier. If we freeze the spins on the edge, as envisaged earlier, the global Z_2 symmetry[4] is broken by the boundary conditions. Notice that such a boundary layer of frozen spins interacts linearly with the neighbouring free spins. It turns out that it is more convenient to introduce a uniform external 'magnetic' field h and to replace the energy by

$$E(\mathcal{C}) = -J \sum_{\langle i,j \rangle} \sigma_i \sigma_j - h \sum_i \sigma_i. \tag{2.13}$$

(The term magnetic field comes from the interaction between a spin $\vec{S}$ and a field $\vec{B}$ proportional to $-\vec{S}.\vec{B}$.) For positive non-zero h, the ground state is indeed the pure state $\sigma_i = +1$, but clearly the Z_2 symmetry is broken and the expectation value (2.12) has no reason to vanish any more. Translation invariance, which is exact if we used a lattice with periodic boundary conditions, implies that the expectation value of σ_i is the same for all sites. The magnetization per site $m = \langle \sigma_i \rangle$ is thus given by

$$m = \frac{1}{N\beta} \frac{\partial \log Z}{\partial h}. \tag{2.14}$$

Clearly, the symmetry is restored when h vanishes:

$$\lim_{h \to 0} m = 0. \tag{2.15}$$

However, we should examine what happens for the thermodynamic limit: $\log Z$ is extensive, i.e., proportional to N. Therefore, the limit of m when N goes to infinity exists a priori. We have seen that, at fixed N, $\lim_{h \to 0} m = 0$, but we may ask whether, in the opposite order of limits,

$$\lim_{h \to 0} \lim_{N \to \infty} m = m_{\mathrm{s}} \neq 0. \tag{2.16}$$

If the right-hand side, m_{s}, of (2.16) does not vanish, the system has a spontaneous magnetization and the symmetry is *spontaneously broken*. In that case, in spite of the vanishing of the symmetry-breaking field h, the system develops a spontaneous magnetization $0 \leq m_{\mathrm{s}} \leq 1$. The remnant of the Z_2 symmetry is, thus, simply the relation

$$\lim_{h \to +0} \lim_{N \to \infty} m = - \lim_{h \to -0} \lim_{N \to \infty} m = m_{\mathrm{s}} \tag{2.17}$$

and the curve $m(h)$ has the characteristic shape of Figs 2.1 and 2.2.

[4] The group Z_2 is the group consisting of the two elements $+1$ and -1 with the usual multiplicative law.

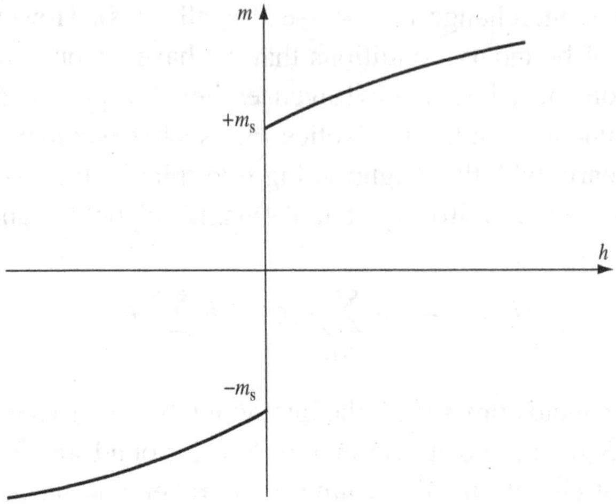

Figure 2.1 An isothermal curve in the (m, h) plane $(T < T_c)$

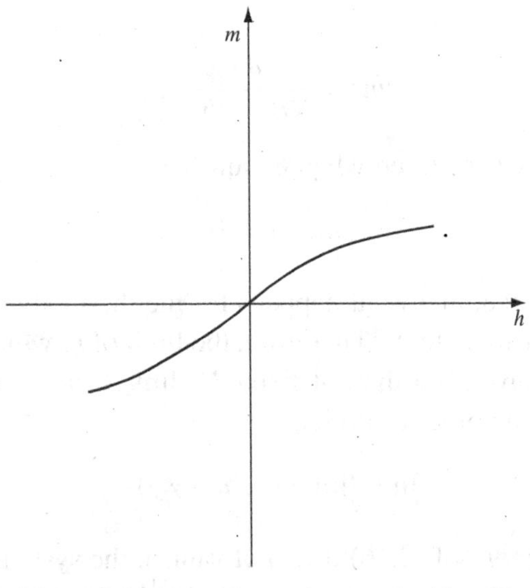

Figure 2.2 An isothermal curve for $T > T_c$

The spontaneous magnetization m_s is the *order parameter* of this transition. It vanishes in the disordered phase (the paramagnetic phase) in which the magnetization vanishes when the applied field goes to zero. It is non-zero in the ordered phase in which the symmetry is broken (ferromagnetism).

I have used the word 'phase' above with several inequivalent meanings that it is important to disentangle.

1. If lowering the temperature, in the absence of any external field, induces a non-zero spontaneous magnetization, below some critical temperature T_c (I will demonstrate below that this may happen) one speaks of a transition from a paramagnetic phase ($T > T_c$) to a ferromagnetic phase ($T < T_c$) (or more generally this is a transition from a disordered phase to an ordered phase driven by the temperature).
2. In the ferromagnetic domain one can induce a transition between the two states of opposite magnetizations by the action of an external magnetic field which goes from a positive to a negative value. The two pure states, ferro $\uparrow$ and ferro $\downarrow$, coexist for $h = 0, T \leq T_c$.

The phase diagram in the (T, h) plane is thus very simple (Fig. 2.3): the coexistence curve of the two phases of opposite magnetizations is the segment of the T-axis $h = 0, 0 \leq T \leq T_c$. Up to a distortion of coordinates, one sees the analogy with a liquid–gas transition. The (m, h) curve of Fig. 2.4 is the equivalent of an isothermal curve in the (p, V) plane for $T < T_c$ while Fig. 2.5 is the equivalent curve for Fig. 2.2. The coexistence segment of Fig. 2.6 is the equivalent of the coexistence curve of a liquid and its vapour in the (p, T) plane. The field h replaces the difference $(p - p_{\text{coexistence}})$ for a fluid: the ferromagnetic $\uparrow$ phase is the equivalent of the liquid, the ferromagnetic $\downarrow$ phase the equivalent of the vapour. (This analogy is further developed in Section 2.6.)

However, up to now, we have admitted without proof the possibility of a spontaneously broken symmetry. The proof will be given in Chapter 4.

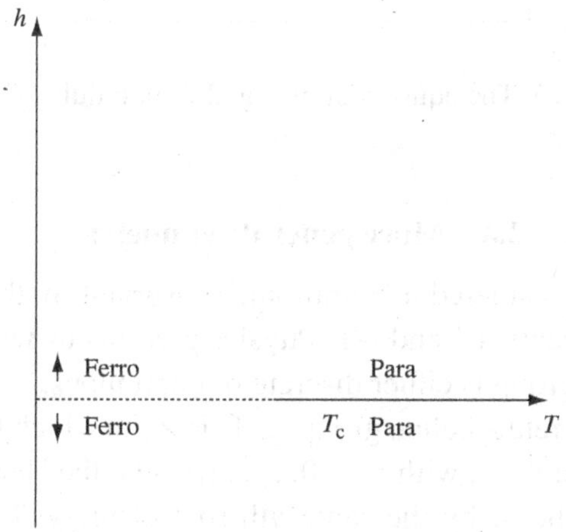

Figure 2.3 The coexistence curve in the (h, T) plane is the segment of the $h = 0$ axis, with $0 < T < T_c$

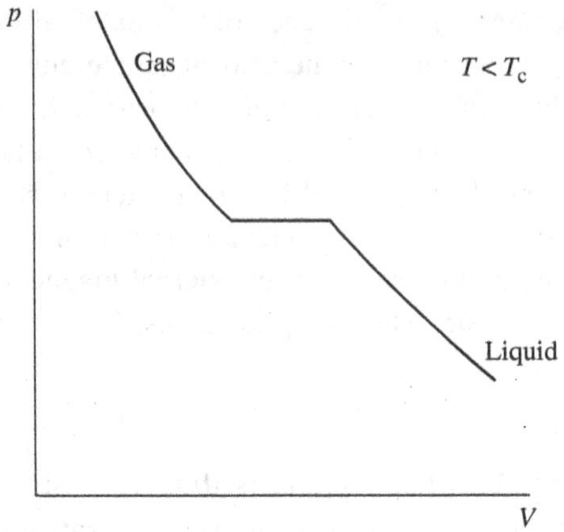

Figure 2.4 The equivalent of Fig. 2.1 for a fluid ($T < T_c$)

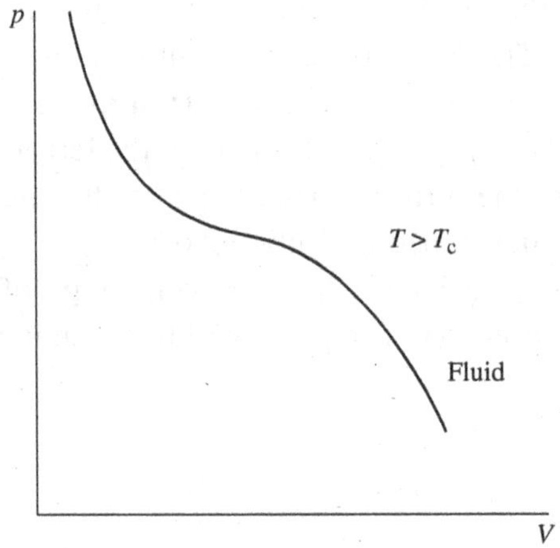

Figure 2.5 The equivalent of Fig. 2.2 for a fluid ($T_c < T$)

2.4 More general symmetries

Up to now we have considered a Hamiltonian invariant by the simplest group Z_2 composed of the elements $+1$ and -1. Physics presents us with other situations in which the symmetry group is either discrete or continuous.

One can have a discrete Abelian group such as Z_n in which the 'spin' varies over the nth roots of unity $e^{2i\pi p/n}$, with $p = 0, \ldots, (n-1)$; the Hamiltonian is invariant if one multiplies all the σ_i by the same nth root of unity. There are now n pure phases that may coexist when the symmetry is broken.

One frequently encounters situations in which the group is the rotation group in n dimensions $O(n)$. The classical 'spins' $\vec{\sigma}_i$ are n-dimensional vectors and the

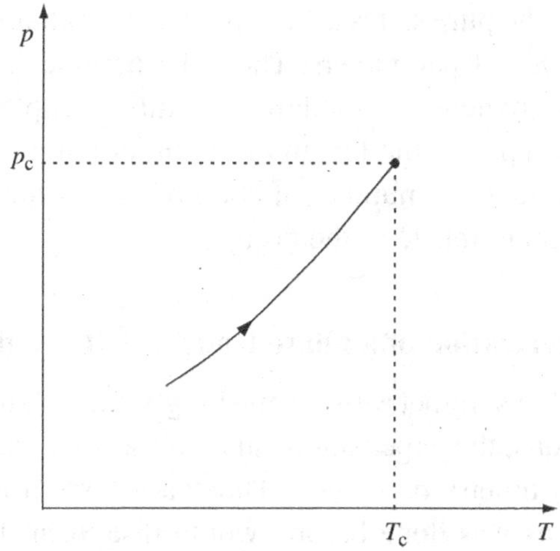

Figure 2.6 Coexistence curve for a fluid, the equivalent of Fig. 2.3

Hamiltonian is invariant under a simultaneous rotation of all these vectors. This number n of components must be distinguished from the dimension d of the space in which the lattice is embedded. For instance, the lattice may be three-dimensional and the spins restricted to vary on a circle. More generally, the spins can vary on the sphere $S^{(n-1)}$. In the high-temperature disordered phase, all the points on the sphere are equivalent and $\langle \vec{\sigma}_i \rangle = 0$. In the low-temperature phase, when the symmetry is spontaneously broken, $\langle \vec{\sigma}_i \rangle \neq 0$. It is the n-component order parameter of the problem. In the phase with broken symmetry, the rotations that leave this vector invariant are still a symmetry of the problem: the subgroup $O(n-1)$ is unbroken. The order parameter is left invariant by a rotation of $O(n-1)$; the various pure states are generated from one of them by the rotations of $O(n)$ that do not belong to the subgroup $O(n-1)$.

More generally, the symmetry group of the Hamiltonian may be a group $\mathcal{G}$; the 'spin' belongs to a representation $\mathcal{R}$ of the group $\mathcal{G}$. In the previous example, $\mathcal{G}$ was the rotation group in n dimensions and $\mathcal{R}$ its vector representation of dimension n. The symmetry may be broken spontaneously to a subgroup $\mathcal{H}$ of $\mathcal{G}$ (in the example $\mathcal{H}$ is the rotation group in $(n-1)$ dimensions that leaves a given vector invariant). A pure state is a vector of the space acted on by the representation $\mathcal{R}$, which is left invariant by $\mathcal{H}$. The action of an element of $\mathcal{G}$ that does not belong to $\mathcal{H}$, or more precisely the action of an element of the quotient $\mathcal{G}/\mathcal{H}$, generates another pure state. For a continuous group $\mathcal{G}$ (a Lie group) possessing p generators, and its subgroup $\mathcal{H}$ having q generators, the manifold of pure states of the low-temperature phase has dimension $p-q$. For instance, if $\mathcal{G} = O(n)$ and $\mathcal{H} = O(n-1)$, $p = n(n-1)/2$, since the generators of $O(n)$ are $n \times n$ antisymmetric matrices,

$q = (n-1)(n-2)/2$. The pure states are the points on a sphere $S^{(n-1)}$, which indeed depends on $p - q = n - 1$ parameters. The order parameter has more parameters than $(p-q)$. In addition there are 'moduli' as in this example in which, in addition to the direction on the sphere, the length of the order parameter introduces an nth degree of freedom. In fact, the number of components of the order parameter also depends on the representation $\mathcal{R}$ of the group.

2.5 Characterization of a phase transition through correlations

If we had not broken the symmetry by a vanishingly small external field, or through the boundary conditions, the expectation value of the spin would vanish at all temperatures from the symmetry $\sigma \to -\sigma$. The characterization of the transition by the order parameter, as was done before, would disappear. In fact, there are systems for which the equivalent of the magnetic field, or of the symmetry-breaking boundary conditions, are not available. For instance, the superfluid transition of liquid helium is characterized by a broken $U(1)$ invariance, i.e., the invariance by a phase redefinition of the creation and annihilation operators for one atom: $\psi(r) \to e^{i\alpha}\psi(r)$ and $\psi^\star(r) \to e^{-i\alpha}\psi^\star(r)$. This symmetry cannot be broken by an infinitesimal source: such a source would indeed create or absorb helium atoms. Nevertheless, there is a transition.

In such cases, one can characterize a broken symmetry by the spatial behaviour of the correlation functions. For instance, in the previous Ising example one can consider the correlation between the spins at spatial points $\vec{r}_1$ and $\vec{r}_2$:

- In the disordered phase ($T > T_c$),

$$\lim_{|\vec{r}_2-\vec{r}_1|\to\infty} \langle\sigma(\vec{r}_1)\sigma(\vec{r}_2)\rangle = 0. \tag{2.18}$$

- However, in the ordered phase ($T < T_c$)

$$\lim_{|\vec{r}_2-\vec{r}_1|\to\infty} \langle\sigma(\vec{r}_1)\sigma(\vec{r}_2)\rangle = m_s^2 \neq 0, \tag{2.19}$$

  and indeed one can prove that the m_s which appears in the r.h.s. of (2.19) is the same as the spontaneous magnetization that was defined in (2.16) through an infinitesimal source.[5]

The existence of a spontaneous magnetization in the sense of (2.19) shows that the effects of a spin extend to large distances. This is better seen on the conditional probability that $\sigma(\vec{r}_2) = +1$ given that $\sigma(\vec{r}_1) = +1$:

$$P\{\sigma(\vec{r}_2) = +1|\sigma(\vec{r}_1) = +1\} = \frac{1}{2}(1 + \langle\sigma(\vec{r}_1)\sigma(\vec{r}_2)\rangle). \tag{2.20}$$

[5] See, e.g., R. Griffiths. In *Phase Transitions and Critical Phenomena*, vol. 1, ed. C. Domb and M. Green. (London: Academic Press, 1972).

Exercise
Prove (2.20).

If m_s vanishes, the spin 1 has no effect on the spin 2; but if m_s is non-zero, the influence propagates far away.

2.6 Phase coexistence, critical points, critical exponents

The discussion, up to now, has been focused on symmetries and the occurrence of spontaneous symmetry breaking. However, if the breaking of a symmetry is well apparent in the paramagnetic to ferromagnetic transition, in many cases there is no a priori symmetry. For instance in the liquid–gas transition, one does not see any symmetry that would be present in the gas and broken in the liquid. The transition is marked by a discontinuity between the density of the liquid ρ_L and that of the gas ρ_G, with a coexistence line in the (p, T) plane ending at the critical point (p_c, T_c) (see Fig. 2.6). However, the solid–fluid line of coexistence, marked by a similar discontinuity in the densities of the two phases, extends to high pressures and high temperatures without a critical point. The liquid–gas transition for $T < T_c$ is thus accompanied by a jump from ρ_G to ρ_L when the pressure increases from a value smaller than $p_{\text{coexistence}}(T)$ to an infinitesimally larger value. This is analogous to the transition for $T < T_c$ between the down and the up spin phase of a ferromagnet when the applied field goes from a negative to a positive value, and the magnetization from $-m_s(T)$ to $+m_s(T)$. At the critical point, this jump vanishes and one defines a characteristic critical exponent β as

$$\rho_G(T) - \rho_L(T) \sim (T_c - T)^\beta \text{ or } m_s(T) \sim (T_c - T)^\beta. \tag{2.21}$$

Above T_c there is only one phase: the fluid phase above the liquid–gas transition, the paramagnetic phase for magnets.[6]

The transitions with jumps are named first-order transitions. Typically, they occur when the free energy, as a function of an 'order parameter', such as the magnetization, or the difference of densities for fluids, jumps from one minimum to a different minimum as some external driving parameter, such as the external field or the pressure, is varied. In such cases, the free energy is continuous at the transition temperature, but its first derivative, the entropy, is discontinuous: hence the name 'first order'. A critical point requires a free energy that varies continuously without jumping from a unique phase, such as the fluid phase or the paramagnetic phase, to a two-phase coexistence, liquid–gas or up–down spins. Such transitions are often called 'second-order transitions', since the free energy, and its first derivative, the

[6] This fluid phase is often called 'supercritical'.

entropy, are continuous at T_c. However, in some cases more derivatives or the free energy may be continuous at T_c, the extreme being the Kosterlitz–Thouless transition[7] for two-dimensional planar ferromagnets, for which the free energy and all of its derivatives are continuous at T_c; however, there is an essential singularity at T_c. In usual cases, unlike the previous one, one defines a critical exponent to characterize the singularity of the free energy, or rather of its second derivative with respect to T, proportional to the specific heat:

$$C(T) \sim (T - T_c)^{-\alpha} \text{ for } T > T_c,$$
$$C(T) \sim (T_c - T)^{-\alpha'} \text{ for } T < T_c. \tag{2.22}$$

There are other frequently used exponents; for instance, the susceptibility, the response of the order parameter to a change of the driving field. In a magnetic case the magnetic susceptibility

$$\chi = \left.\frac{\partial m}{\partial h}\right|_{h=0} \tag{2.23}$$

diverges at T_c: a vanishingly small field is sufficient to induce a large magnetization in the paramagnetic phase. Its equivalent for fluids is the isothermal compressibility. The divergence at T_c leads to more exponents:

$$\chi(T) \sim (T - T_c)^{-\gamma} \text{ for } T > T_c,$$
$$\chi(T) \sim (T_c - T)^{-\gamma'} \text{ for } T < T_c. \tag{2.24}$$

The critical isotherm in the (p, V) plane, or in the (h, m) plane at $T = T_c$ leads to the exponent δ

$$h \sim m^{\delta} \text{ or } p - p_c \sim (\rho_L - \rho_G)^{\delta}. \tag{2.25}$$

The last exponents are relative to the correlation function between widely spaced points in the sample. In the magnetic case one considers the correlation function

$$G(\vec{r}, T) = \langle s(\vec{r}) s(0) \rangle. \tag{2.26}$$

When $|\vec{r}|$ is large, this correlation function falls off exponentially with distance

$$G(\vec{r}, T) \sim |_{r \to \infty}\, \mathrm{e}^{-r/\xi(T)} \tag{2.27}$$

up to some power of r. The correlation length $\xi(T)$ becomes very large as one gets closer to the critical point[8] and one defines the exponents

[7] J. M. Kosterlitz and D. J. Thouless, *J. Phys. C: Solid State Phys.*, **6** (1973) L97.

[8] For fluids this is manifested by the phenomenon of critical opalescence, which shows a correlation length of the order of the wavelength of light, three orders of magnitude above the typical intermolecular distance.

$$\chi(T) \sim (T - T_c)^{-\nu} \text{ for } T > T_c,$$
$$\chi(T) \sim (T_c - T)^{-\nu'} \text{ for } T < T_c. \tag{2.28}$$

Last, but not least, at T_c itself the correlation function falls off as a power of distance;

$$G(\vec{r}, T) \sim |_{r\to\infty} \frac{1}{r^{d-2+\eta}}, \tag{2.29}$$

in which d is the space dimension; this curious definition of η comes in fact from the Fourier transform $\tilde{G}(\vec{p}, T)$, which diverges at low momentum as $1/p^{2-\eta}$. The exponents η and ν will play an important rôle in later discussions.

The understanding of the nine critical exponents that I have defined up to now has been a central problem of the theory of phase transitions. We shall see that they are not independent (scaling laws) and that they are *universal*, i.e., independent of the specific system under consideration, given the nature of the broken symmetry.

I have insisted that the liquid–gas transition has no symmetry; however, it follows from the subsequent theory that, in the vicinity of the critical point, its singularities are identical to those of an Ising ferromagnet. Therefore, we shall focus in the following on broken symmetries.

Nature presents us with a large number of phase coexistence lines ending at a critical point: in addition to the liquid–gas coexistence, and to the Curie transition from paramagnets to ferromagnets, binary mixtures of liquids or solid alloys present a critical point above which the two species are uniformly mixed. Let us recall that ^{4}He presents two liquid phases at low temperatures, a normal one and a superfluid one, with a critical point at $T_c \simeq 2.17$ K. ^{3}He has two different superconducting phases (with a p-wave pairing of two atoms) in the millikelvin range. The phase diagram of superconductors in an external magnetic field is also very rich. If the interaction between spins tends to make neighbouring spins antiparallel, one can have a low temperature antiferromagnetic phase in which the magnetization alternates in sign from site to site, with a critical point at the Néel temperature. An external uniform field is not conjugate to the order parameter, which is a staggered magnetization. Its effect is to reduce the magnetization on one sublattice and increase it on the other one. Therefore, there is a line of critical Néel points as the applied field varies. If one keeps increasing the field, the line ends at a *tricritical point*. Above the tricritical point the transition from an antiferromagnet to a paramagnet becomes first order. The theory of tricriticality is very similar in essence to the one that will be developed here, but I will not discuss it in this book.

3

Examples of physical situations modelled by the Ising model

The examples are numerous and we shall limit ourselves to a few classical examples.

3.1 Heisenberg's exchange forces

In spite of the fact that the interactions between electrons in a solid, as well as the forces exerted by the ions on the electrons, are essentially of Coulomb origin, the necessity of an interaction between the magnetic moments of the electrons had been recognized long ago. First of all, a classical result obtained by Niels Bohr and Hendrika van Leeuwen demonstrates that magnetism is a purely quantum phenomenon: in classical statistical mechanics an applied field does not produce any magnetization.

Exercise

Prove the Bohr–van Leeuwen theorem. (Hint: the effect of the applied field $\vec{B}$ is to replace every electronic momentum $\vec{p}$ by $\vec{p} - e\vec{A}$ in which $\vec{B} = \vec{\nabla} \wedge \vec{A}$.)

Indeed, Langevin had foreseen that there should be microscopic magnetic moments long before the discovery of spin. In the paramagnetic phase, these moments are randomly oriented and the average magnetization vanishes in the absence of an external field. On the other hand, in the ferromagnetic phase these moments act coherently. Understanding the origin of this coherent effect was a major puzzle, solved by Heisenberg in 1932. Such magnetic moments interact by a dipole–dipole interaction, but this interaction is much too weak to explain the magnetization of metals, such as iron oxides, at room temperature. Heisenberg understood that an effective interaction between the spins of the electrons, thousands of times larger than the dipolar interaction, was induced by the Pauli principle, i.e., the antisymmetrization of the electronic states.

Let us consider the simple problem of two electrons located at $\vec{r}_1$ and $\vec{r}_2$ in the field of two fixed ions located at $\vec{R}_1$ and $\vec{R}_2$. The Hamiltonian has the form

$$H = H_0 + \frac{e^2}{|\vec{r}_1 - \vec{r}_2|}, \tag{3.1}$$

with

$$H_0 = h(1) + h(2), \tag{3.2}$$

in which $h(1)$ describes the kinetic energy and the effect of the two ions on the first electron; and the same for $h(2)$. Let us denote by $\varphi_\lambda(\vec{r})$ the eigenfunctions of h and by ϵ_λ the corresponding eigenvalues.

The functions $\varphi_\lambda(\vec{r}_1)\varphi_\mu(\vec{r}_2)$ and $\varphi_\lambda(\vec{r}_2)\varphi_\mu(\vec{r}_1)$ are eigenfunctions of H_0 for the same eigenvalue $\epsilon_\lambda + \epsilon_\mu$. Any linear combination of these two functions is also an eigenfunction of H_0 for that same eigenvalue, in particular

$$\Phi_\mathrm{s}(\vec{r}_1, \vec{r}_2) = \frac{1}{\sqrt{2}}(\varphi_\lambda(\vec{r}_1)\varphi_\mu(\vec{r}_2) + \varphi_\lambda(\vec{r}_2)\varphi_\mu(\vec{r}_1)) \tag{3.3}$$

and

$$\Phi_\mathrm{a}(\vec{r}_1, \vec{r}_2) = \frac{1}{\sqrt{2}}(\varphi_\lambda(\vec{r}_1)\varphi_\mu(\vec{r}_2) - \varphi_\lambda(\vec{r}_2)\varphi_\mu(\vec{r}_1)), \tag{3.4}$$

which are respectively symmetric and antisymmetric under the exchange of the positions of the two electrons. However, Pauli's principle implies that we must retain only states that are antisymmetric by exchange of all the quantum numbers of the two electrons, namely their positions *and their spins.*

The four possible spin states of the pair of electrons $|\pm, \pm\rangle$ may be rearranged into states of total spin equal to one or to zero:

- Spin 1 triplet states,

$$\begin{aligned} |S = 1, S_z = 1\rangle &= |+, +\rangle, \\ |S = 1, S_z = 0\rangle &= \tfrac{1}{\sqrt{2}}(|+, -\rangle + |-, +\rangle), \\ |S = 1, S_z = -1\rangle &= |-, -\rangle. \end{aligned} \tag{3.5}$$

- And one spin 0 singlet state,

$$|S = 0, S_z = 0\rangle = \frac{1}{\sqrt{2}}(|+, -\rangle - |-, +\rangle). \tag{3.6}$$

The spin 1 states are symmetric under spin exchange, whereas the spin 0 state is antisymmetric. One can thus construct four eigenstates of H_0 for the energy $\epsilon_\lambda + \epsilon_\mu$ satisfying Pauli's principle, namely

$$\begin{aligned} &\Phi_\mathrm{s}(\vec{r}_1, \vec{r}_2)\ |S = 0, S_z = 0\rangle, \\ &\Phi_\mathrm{a}(\vec{r}_1, \vec{r}_2)\ |S = 1, S_z = +1\rangle \text{ or } |S = 1, S_z = 0\rangle \text{ or } |S = 1, S_z = -1\rangle. \end{aligned} \tag{3.7}$$

Up to now we have neglected the Coulomb repulsion between the two electrons. Let us show that this repulsion has the effect of partially lifting the degeneracy between the four eigenstates of H_0. If we compute the effect of this repulsion to first order in perturbation theory, the energy will take the values $\epsilon_\lambda + \epsilon_\mu + \delta\epsilon_\mathrm{T}$ three-fold degenerate, or $\epsilon_\lambda + \epsilon_\mu + \delta\epsilon_\mathrm{S}$ with

$$\begin{aligned}\delta\epsilon_\mathrm{T} &= \left\langle \Phi_\mathrm{a} \left| \frac{e^2}{|\vec{r}_1 - \vec{r}_2|} \right| \Phi_\mathrm{a} \right\rangle, \\ \delta\epsilon_\mathrm{S} &= \left\langle \Phi_\mathrm{s} \left| \frac{e^2}{|\vec{r}_1 - \vec{r}_2|} \right| \Phi_\mathrm{s} \right\rangle\end{aligned} \tag{3.8}$$

and consequently

$$\delta\epsilon_\mathrm{T} - \delta\epsilon_\mathrm{S} = -2 \int \mathrm{d}\vec{r}_1 \mathrm{d}\vec{r}_2 \varphi_\lambda^\star(\vec{r}_1)\varphi_\mu^\star(\vec{r}_2) \frac{e^2}{|\vec{r}_1 - \vec{r}_2|} \varphi_\mu(\vec{r}_1)\varphi_\lambda(\vec{r}_2). \tag{3.9}$$

The r.h.s. of (3.9) is called the 'exchange integral': it is manifestly a quantum effect in which one cannot say which electron is in which state. This result is quite remarkable: the interactions that enter in this calculation are purely Coulombian, spin independent, and yet the Pauli principle yields an effective spin-dependent interaction.

3.2 Heisenberg and Ising Hamiltonians

This difference of energy between the triplet and singlet states, in a physical situation in which one can neglect the transition to other excited states, may be reproduced by an effective Hamiltonian that possesses the right energy splitting and degeneracies. The Hamiltonian

$$H_\mathrm{eff} = -\frac{J}{2}(\vec{S}_1 + \vec{S}_2)^2 = -J\left(\vec{S}_1 \cdot \vec{S}_2 + \frac{3}{4}\right) \tag{3.10}$$

($\hbar = 1$) has indeed two energy levels: a singlet state of zero energy and a triplet state of energy $-J$. If the exchange integral J is positive, the energy is lower for two parallel spins and this interaction may potentially lead to a situation in which the spins of a many-body system become ordered at low temperature.

In fact, the situation is more delicate and the previous discussion is quite insufficient. If one considers the two electrons of a hydrogen molecule H_2, the exchange integral is negative. One can even show that the ground state has spin zero when the forces are spin independent. In a solid that possesses localized magnetic moments, the mechanism of interaction between these moments is mediated through an indirect exchange via the delocalized conduction electrons. A more serious discussion would involve a real solid-state physics course.

Returning to the exchange integral, the exponential fall-off of the wave functions $\varphi_\lambda(\vec{r})$ and $\varphi_\mu(\vec{r})$ with distance implies a rapid decrease of this integral when the distances between the ions $|\vec{R}_1 - \vec{R}_2|$ increases. This leads to the Heisenberg model in which one considers localized magnetic moments, i.e., spins, located at the nodes of a periodic crystalline lattice; the exchange integral $J(\vec{r} - \vec{r}')$ between two sites $\vec{r}$ and $\vec{r}'$ is rapidly decreasing with their distance and one can assume that the interactions are limited to nearest neighbours. These considerations led Heisenberg to propose the Hamiltonian

$$H_{\text{Heis.}} = -J \sum_{\langle i,j \rangle} \vec{S}_i \cdot \vec{S}_j. \tag{3.11}$$

Understanding the properties of a large system of spins from this model raises many problems. First of all, the diagonalization of this Hamiltonian is not simple: $\vec{S}_i \cdot \vec{S}_j$ does not commute with $\vec{S}_i \cdot \vec{S}_k$.

Exercise

Take three sites and a spin $\frac{1}{2}$ at each site. Find the eigenvalues of $H = -J(\vec{S}_1 \cdot \vec{S}_2 + \vec{S}_2 \cdot \vec{S}_3)$ and their degeneracies. Repeat for the case where the three sites are on a circle, i.e., $H = -J(\vec{S}_1 \cdot \vec{S}_2 + \vec{S}_2 \cdot \vec{S}_3 + \vec{S}_3 \cdot \vec{S}_1)$.

For a one-dimensional chain, with nearest-neighbour interactions, H. Bethe[1] has found the eigenstates of this Hamiltonian through the celebrated 'Bethe ansatz', which has been the model for finding solutions of numerous integrable models. However, no method for diagonalizing this Hamiltonian in dimensions larger than one has ever been found and it seems out of reach.

Furthermore, even when all the eigenvalues of the Hamiltonian are known, computing the partition function in the thermodynamic limit is still a formidable problem, as we shall see below with the Ising model.

To simplify the algebra, Lenz and Ising have introduced a model in which the energy $\vec{S}_1 \cdot \vec{S}_2$ of two neighbouring spins is replaced arbitrarily by $S_1^z S_2^z$. One then replaces the Heisenberg Hamiltonian (3.11) with the Ising Hamiltonian,

$$H_{\text{Ising}} = -J \sum_{\langle i,j \rangle} S_i^z S_j^z. \tag{3.12}$$

The simplification is considerable, since all the terms of the Hamiltonian commute and the eigenvalues of H_{Ising} are specified by assigning to each site one of the two eigenvalues of $S_i^z = \frac{1}{2}\sigma_i$ with $\sigma_i = \pm 1$. An assignment of the N eigenvalues σ_i for the whole lattice gives a configuration $\mathcal{C}$ with energy

[1] H. Bethe, *Zeitschrift für Physik A*, **71** (1931) 205–26.

$$E(\mathcal{C}) = -\frac{J}{4}\sum_{\langle i,j\rangle}\sigma_i\sigma_j. \tag{3.13}$$

In the presence of a uniform external field h along the same z-axis the energy of $\mathcal{C}$ is

$$E(\mathcal{C}) = -J\sum_{\langle i,j\rangle}\sigma_i\sigma_j - h\sum_i \sigma_i \tag{3.14}$$

(J has been changed to $4J$). Of course, it remains necessary to compute the partition function by summing over the 2^N configurations $\mathcal{C}$, a problem that we shall consider in the next chapter.

The simplification, which consists of neglecting the components S_x and S_y of every spin, is not innocent: the physics is not the same in many respects for the Heisenberg and Ising models. However, there are in nature ferromagnets in which the crystalline anisotropies allow the spins to point in one single direction (uniaxial ferromagnets); in such systems the Ising model is physically relevant.

3.3 Lattice gas

Consider molecules at thermal equilibrium in a box of volume V, in a range of temperatures and densities, in which one can ignore the quantum effects. If we treat this system in the grand canonical ensemble, the partition function is

$$Z = \sum_N \frac{e^{\alpha N}}{h^{3N}N!}\int d\vec{r}_1 d\vec{p}_1\cdots d\vec{r}_N d\vec{p}_N e^{-\beta\left[\sum_1^N \frac{\vec{p}_i^2}{2m}+U(\vec{r}_1\cdots\vec{r}_N)\right]}, \tag{3.15}$$

in which $U(\vec{r}_1\cdots\vec{r}_N)$ is the interaction energy of N molecules and α is the Lagrange parameter related to the chemical potential by $\alpha/\beta = \mu$. Integrating over the momenta, one obtains

$$Z = \sum_N \frac{z^N}{N!}Q_N \tag{3.16}$$

with

$$z = \frac{e^{\alpha}}{h^3}(2\pi mkT)^{3/2} \tag{3.17}$$

and

$$Q_N = \int d\vec{r}_1\cdots d\vec{r}_N e^{-\beta U(\vec{r}_1\cdots\vec{r}_N)}. \tag{3.18}$$

To compute approximately this integral Q_N, one discretizes the volume V with a periodic lattice of $\mathcal{N}$ points. The total volume V is the product of the volume v of the elementary cell by $\mathcal{N}$ (for a cubic lattice). The molecules are located at the sites of this lattice. The interaction energy U contains a repulsion at short distance between the molecules, which we take into account by forbidding the occupation of a site of the lattice by more than one molecule. The sum over N in (3.16) is thus limited to $N \leq \mathcal{N}$.

The interaction energy also contains a short-range attraction between pairs of molecules. This may be modelled schematically by assigning an energy $-\epsilon$ to a pair of molecules that are nearest neighbours on the lattice; since the interaction is short ranged, we assume no potential energy for molecules which are not nearest neighbours. The total configuration energy is obtained by multiplying $-\epsilon$ by the number of pairs of nearest neighbours.

A configuration is a choice of N sites among the $\mathcal{N}$. Once this choice is made, there are $N!$ ways of distributing the molecules over the N sites. It is convenient to define the *occupation number* of a site i, i.e., a number n_i which is equal to 1 if the site i is occupied and to 0 if it is empty. These n_i are not completely independent, since

$$\sum_{1}^{\mathcal{N}} n_i = N. \tag{3.19}$$

In terms of these occupation numbers, the interaction energy of a configuration C of the N molecules is given by

$$U(C) = -\epsilon \sum_{\langle i,j \rangle} n_i n_j. \tag{3.20}$$

Since this energy is $N!$ times degenerate, one has then in this discretized formulation

$$Q_N = N! \sum_{C} \mathrm{e}^{-\beta U(C)}. \tag{3.21}$$

The constraint (3.19) makes the computation of the canonical partition function more difficult, since the configurations cannot be described by an independent assignment of the n_i to the whole lattice. However, the grand-canonical partition function is given simply by

$$Z = \sum_{C} \mathrm{e}^{(n_1 + \cdots + n_{\mathcal{N}}) \log z + \beta\epsilon \sum_{\langle i,j \rangle} n_i n_j}, \tag{3.22}$$

in which C is an assignment of 0 and 1 independently for each site over the whole lattice of $\mathcal{N}$ sites.

Exercise
Prove (3.22).

To complete the identification with the Ising model one introduces a 'spin' for each occupation number:

$$n_i = \frac{1}{2}(1 + \sigma_i) \quad \text{with } \sigma_i = \pm 1. \tag{3.23}$$

This leads to

$$Z = \sum_{\mathcal{C}} \mathrm{e}^{-\beta E(\mathcal{C})}, \tag{3.24}$$

with

$$E(\mathcal{C}) = -J \sum_{\langle i,j \rangle} \sigma_i \sigma_j - h \sum_i \sigma_i + \text{constant}, \tag{3.25}$$

in which

$$J = \frac{\epsilon}{4} \quad \text{and} \quad h = \frac{\log z}{2\beta} + \frac{c\epsilon}{4} \tag{3.26}$$

(c is the number of neighbours of a given site).

If the Ising model presents a ferromagnetic order at temperatures $T < T_{\rm c}$, it means that a change of h from $+0$ to -0 implies a transition between two phases of opposite magnetizations, which coexist along the line $h = 0$. How can one translate this into molecular language? The magnetization $\langle \sigma \rangle$ is related to the mean occupation of the sites. A jump of the magnetization from $+m_{\rm s}$ to $-m_{\rm s}$ is thus a discontinuity that is equivalent to the mean number of molecules per unit volume from a dense liquid phase to a gaseous phase. These two phases are spatially homogeneous ($\langle n_i \rangle$ is the same for all sites). The coexistence curve of the two phases in the plane (chemical potential versus temperature) is given by $h = 0$, $T \leq T_{\rm c}$, i.e.,

$$\mu + \frac{c\epsilon}{2} + kT \log \left(\frac{2\pi mkT}{h^2} \right)^{3/2} = 0. \tag{3.27}$$

3.4 More examples

Numerous order–disorder transitions may be described in an analogous way. For instance, solid binary alloys, such as copper–zinc, may be ordered at low temperature (the atoms of Cu are at the nodes of a cubic lattice and the Zn atoms are at the centres of these cubes, making this also a cubic lattice). At higher temperatures, substitutions of a copper atom by a zinc atom occur until the transition beyond which the probabilities of finding a copper atom on both sublattices are equal. Again, a 'spin' σ_i will characterize the nature of the atom located at the site i. Formal neural networks have been modelled in similar ways: the neuron has a 'spin', which takes one value if the neuron 'fires', and another one if it is at rest. The neurons are coupled by 'synapses' but some are excitatory, and others inhibitory.

Therefore, one needs to introduce coupling $J_{i,j}$ with different signs. One enters the domain of 'spin glasses', which goes beyond the scope of this book.

3.5 A first connection with field theory

Let us consider a simple relativistic quantum field theory, for instance a neutral self-coupled scalar field φ for which the action is (in units $c = 1$)

$$S = \int \mathrm{d}t \mathrm{d}\vec{x} \left[\frac{1}{2} \left(\frac{\partial \varphi}{\partial t} \right)^2 - \frac{1}{2} (\nabla \varphi)^2 - V(\varphi) \right], \tag{3.28}$$

in which $V(\varphi)$ contains both mass terms and interactions. In Feynman's path integral formulation the amplitude associated to a physical process is the sum over all 'configurations' of the field, weighted by the complex number $\mathrm{e}^{\mathrm{i}S/\hbar}$.

An analytic extension to imaginary time (Wick rotation) can be performed:

$$t \to -\mathrm{i}x_4 \qquad\qquad \frac{\partial \varphi}{\partial t} \to \mathrm{i}\frac{\partial \varphi}{\partial x_4}, \tag{3.29}$$

in which

$$\mathrm{e}^{\mathrm{i}S/\hbar} \to \mathrm{e}^{-A} \tag{3.30}$$

with the 'Euclidean' action,

$$A = \frac{1}{\hbar} \int \mathrm{d}^4 x \left[\frac{1}{2} \sum_{a=1}^{4} \left(\frac{\partial \varphi}{\partial x_a} \right)^2 + V(\varphi) \right]. \tag{3.31}$$

The continuation of the Euclidean theory to real time allows one to recover the real physics. In this Euclidean theory we have again a positive Boltzmann weight, e^{-A}, with A in the rôle of energy and $\hbar$ replacing kT: the quantum fluctuations are replaced by thermal fluctuations. Note that at this stage the four-vector x_a varies continuously over R^4.

Computing Feynman's sum over histories of e^{-A} is a sum over all the 'configurations' of the field $\varphi(\mathbf{x})$, with prescribed boundary conditions. It is a functional integral, an integral in which there are an infinite number of integration variables, namely the value of $\varphi(\mathbf{x})$ at every $\mathbf{x} \in R^4$. Apart from trivial cases, such as free fields, which lead to Gaussian integrals, one does not know how to compute such sums. Perturbation theory, in which one expands the terms of degree higher than two in the action, generates ultraviolet divergences, i.e., short distance singularities. I shall discuss later these points in detail, but at this stage we simply note that the theory is ill-defined until it is *regularized* by some cut-off Λ on large momenta, a scale much larger than the physical masses and momenta. If, instead, one works in position space, a momentum cut-off has the effect of killing the spatial variations of the field with wavelength smaller than Λ^{-1}. One can thus regularize the theory by the introduction of an elementary length $a \simeq \Lambda^{-1}$. One then replaces the space–time continuum by a four-dimensional periodic lattice with lattice spacing a. This regularization is of constant use in Monte-Carlo numerical simulations of quantum field theories. We shall meet later other regularizations, which are more convenient for analytic calculations, but let us examine further the effect of this lattice regularization.

The four-dimensional space–time is thus replaced by a hypercubic lattice, with lattice spacing a. The following replacements follow this discretization of space–time:

$$\varphi(\mathbf{x}) \rightarrow \varphi_i = \varphi(x_i), \tag{3.32}$$

in which the x_i are the nodes of the lattice,

$$\int \mathrm{d}^4 x V(\varphi(\mathbf{x})) \rightarrow a^4 \sum_i V(\varphi_i), \tag{3.33}$$

the field derivative by

$$\frac{\partial \varphi}{\partial x_\mu} \rightarrow a^{-1}[\varphi(\mathbf{x} + a e_\mu) - \varphi(\mathbf{x})], \tag{3.34}$$

in which e_μ is the unit vector in the direction μ; the kinetic energy is thus replaced by

$$\begin{aligned} \int \mathrm{d}^4 x \sum_{a=1}^{4} \left(\frac{\partial \varphi}{\partial x_a}\right)^2 &\rightarrow a^2 \sum_i \sum_{\mu=1}^{4} [\varphi(x_i + a e_\mu) - \varphi(x_i)]^2 \\ &= 8a^2 \sum_i \varphi_i^2 - 2a^2 \sum_{\langle i,j \rangle} \varphi_i \varphi_j, \end{aligned} \tag{3.35}$$

in which, as usual, the bracket $\langle i, j \rangle$ restricts the sum to nearest neighbours. The functional integral

$$Z = \int D\varphi \mathrm{e}^{-A} \tag{3.36}$$

is replaced by

$$Z = \int \prod \mathrm{d}\tau_i \ \mathrm{e}^{\frac{a^2}{\hbar} \sum_{\langle i,j \rangle} \varphi_i \varphi_j} \tag{3.37}$$

with a measure

$$\mathrm{d}\tau_i = \mathrm{e}^{-\frac{1}{\hbar}[4a^2 \varphi_i^2 + a^4 V(\varphi_i)]} \mathrm{d}\varphi_i. \tag{3.38}$$

The problem looks like an Ising model, except that the measure $\mathrm{d}\tau$ is not a delta function centred at ± 1, but some other probability distribution.

Note that it is only the coupling between neighbours (3.37) that introduces correlations between distinct lattice sites; without it the random variables φ_i would be all independent. Therefore, the propagation is induced by the (3.37) coupling between neighbours. In general the correlations between distinct points of the lattice fall off exponentially with the distance, with a characteristic length that is comparable to the lattice spacing a. This is a problem: remember that the spatial exponential decrease of the correlations in quantum field theory is characterized by a length scale proportional to the inverse of the masses of the particles. For instance, in a scalar free field theory

$$\begin{aligned}\langle\varphi(\mathbf{x})\varphi(\mathbf{y})\rangle &= \int \mathrm{d}^4 p \frac{\mathrm{e}^{i\mathbf{p}\cdot(\mathbf{x}-\mathbf{y})}}{\mathbf{p}^2+m^2} \\ &\simeq_{|\mathbf{x}-\mathbf{y}|\to\infty} |\mathbf{x}-\mathbf{y}|^{-3/2}\mathrm{e}^{-m|\mathbf{x}-\mathbf{y}|}\end{aligned} \tag{3.39}$$

falls off with a characteristic length $\xi = 1/m$.

Exercise

Verify (3.39); it is convenient to use the identity $1/A = \int_0^\infty \mathrm{d}t\mathrm{e}^{-At}$.

Therefore, a fall-off on the lattice with length scale of order a would correspond to masses of order $\Lambda = a^{-1}$, whereas the cut-off Λ only makes sense if it is much larger than the masses. Consequently, a sensible 'renormalized' theory, with $\Lambda \gg m$, exists only in a situation in which the correlation length is much larger than a. With such short-range interactions this occurs only in the vicinity of a *critical point* at which the correlation length goes to infinity. We shall see that such critical points occur only when a symmetry is spontaneously broken.

The *renormalized* theory, which is the limit when the cut-off Λ goes to infinity, or the lattice spacing a to zero, exists only in the vicinity of a point where a symmetry, such as $\varphi \to -\varphi$, is spontaneously broken. Note again that this regularized field theory differs only from the Ising model in the measure

$$\mathrm{d}\tau_{\text{Ising}} = \delta(\varphi^2 - 1)\mathrm{d}\varphi, \tag{3.40}$$

whereas

$$\mathrm{d}\tau_{\text{field theory}} = \mathrm{e}^{-\frac{1}{\hbar}[4a^2\varphi^2 + a^4 V(\varphi)]}\mathrm{d}\varphi. \tag{3.41}$$

If $V(\varphi)$ contains a negative 'bare mass', as in

$$V(\varphi) = -\frac{1}{2}m_0^2\varphi^2 + \frac{g}{4}\varphi^4, \tag{3.42}$$

the field measure is a peak around the two values $\varphi = \pm\sqrt{\frac{a^2}{g}(m_0^2a^2 - 8)}$ and it is not so different from the Ising measure. A much more elaborate connection between the two models will be the object of the last chapters.

4

A few results for the Ising model

If we were dealing with a quantum model, such as the Heisenberg model, the calculation of the partition function would first require us to find all the eigenvalues of the Hamiltonian. In the case of the Ising model, the eigenvalues are given; they are simply the energies

$$E(\mathcal{C}) = -J \sum_{\langle i,j \rangle} \sigma_i \sigma_j - h \sum_i \sigma_i \tag{4.1}$$

with $\sigma_i = \pm 1$. In spite of this simplicity one must realize that $\sum_{\mathcal{C}} \mathrm{e}^{-\beta E(\mathcal{C})}$ is a sum over 2^N terms. To describe a collective phenomenon, such as a phase transition, in which the correlations extend to large distances, one needs to deal with a lattice containing a sizeable number of points in each spatial direction. A three-dimensional lattice of, say, 20^3 points requires a sum over 2^{8000} configurations, which is, of course, out of reach; the largest lattice for which a complete sum over all configurations has been performed consists of 4^3 points, i.e., about 10^{19} configurations. An appropriate use of symmetries may help to reduce this number slightly but this is clearly the limit, and yet such a small lattice cannot really be a guide to understanding long-range correlations. The only numerical methods that have been used with success are stochastic samplings of configurations, such as Monte-Carlo algorithms, or diagonalization of transfer matrices, which I shall define in the next section.

4.1 One-dimensional Ising model: transfer matrix

The N sites are regularly spaced along a chain, with spacing a. The model, as defined, is easy to solve for all kinds of boundary conditions, but the calculations are even simpler if the chain is closed as if the sites were on a circle, i.e., if we use periodic boundary conditions (p.b.c.):

- Open chain:

$$E(\mathcal{C}) = -J \sum_{1}^{N-1} \sigma_i \sigma_{i+1} - h \sum_{1}^{N} \sigma_i, \tag{4.2}$$

- Periodic boundary conditions:

$$E(\mathcal{C}) = -J \sum_{1}^{N} \sigma_i \sigma_{i+1} - h \sum_{1}^{N} \sigma_i, \tag{4.3}$$

with the identification $\sigma_{N+1} = \sigma_1$.

Clearly, the addition of one extra term in the energy has no effect on the thermodynamic limit (it is easy here to calculate the free energy for open or periodic boundary conditions and verify this statement).

Let us define

$$T(\sigma, \sigma') = \mathrm{e}^{\beta[J\sigma\sigma' + \frac{1}{2}h(\sigma+\sigma')]}. \tag{4.4}$$

Then for p.b.c. one verifies immediately that

$$\mathrm{e}^{-\beta E(\mathcal{C})} = T(\sigma_1, \sigma_2) T(\sigma_2, \sigma_3) \ldots T(\sigma_{N-1}, \sigma_N) T(\sigma_N, \sigma_1). \tag{4.5}$$

If one displays the four possible values of $T(\sigma, \sigma')$ when $\sigma, \sigma' = \pm 1$, one reads:

$$\mathcal{T} = \begin{pmatrix} \mathrm{e}^{\beta(J+h)} & \mathrm{e}^{-\beta J} \\ \mathrm{e}^{-\beta J} & \mathrm{e}^{\beta(J-h)} \end{pmatrix}, \tag{4.6}$$

in which the rows are labelled by the successive values $+1$ and -1 of σ, and the columns by the values of σ'.

Summing over all configurations means summing over $\sigma_1 = \pm 1, \sigma_2 = \pm 1, \ldots, \sigma_N = \pm 1$. Let us take the Boltzmann factor as written in (4.5) and sum over σ_2: it occurs only in the product of the first two factors $T(\sigma_1, \sigma_2) T(\sigma_2, \sigma_3)$. If we return to the matrix expression (4.6), the first σ_2 is an index of a column, the second of a row: therefore, the sum over σ_2 gives simply a product of the two matrices:

$$\sum_{\sigma_2} T(\sigma_1, \sigma_2) T(\sigma_2, \sigma_3) = \mathcal{T}^2(\sigma_1, \sigma_3), \tag{4.7}$$

in which $\mathcal{T}^2$ is the square of the matrix (4.6).

Repeating this argument for the summation over $\sigma_3, \ldots, \sigma_N$ one arrives at

$$\sum_{\sigma_2, \ldots, \sigma_N} \mathrm{e}^{-\beta E(\mathcal{C})} = \mathcal{T}^N(\sigma_1, \sigma_1). \tag{4.8}$$

One last sum remains:

$$Z = \sum_{\sigma_1} \mathcal{T}^N(\sigma_1, \sigma_1), \tag{4.9}$$

the sum over the diagonal elements of the matrix $\mathcal{T}^N$. Therefore, we end up with

$$Z = \mathrm{Tr}\, \mathcal{T}^{\mathrm{N}}. \tag{4.10}$$

The trace of a matrix is the sum of its diagonal elements, it is also the sum of its eigenvalues. We have thus to find the two eigenvalues of the matrix $\mathcal{T}^N$ but these eigenvalues are simply the Nth powers of the eigenvalues of $\mathcal{T}$. Therefore, we end up with

$$Z = \lambda_1^N + \lambda_2^N, \tag{4.11}$$

in which λ_1 and λ_2 are the two roots of the quadratic equation

$$\lambda^2 - \lambda(\mathrm{e}^{\beta(J+h)} + \mathrm{e}^{\beta(J-h)}) + \mathrm{e}^{2\beta J} - \mathrm{e}^{-2\beta J} = 0, \tag{4.12}$$

namely

$$\lambda_1, \lambda_2 = \mathrm{e}^{\beta J} \cosh \beta h \pm \sqrt{\mathrm{e}^{2\beta J} \sinh^2 \beta h + \mathrm{e}^{-2\beta J}}. \tag{4.13}$$

By convention, we will denote by λ_1 the larger of the two eigenvalues (plus sign before the square root).

Then the free energy is given by

$$\begin{aligned} F &= -kT \log Z = -kT \log(\lambda_1^N + \lambda_2^N), \\ &= -NkT \log \lambda_1 - kT \log\left(1 + \left(\frac{\lambda_2}{\lambda_1}\right)^N\right). \end{aligned} \tag{4.14}$$

In the thermodynamic limit in which N goes to infinity, the last term is exponentially small as long as $\lambda_2/\lambda_1 < 1$, which is always true (except at $h = 0, T = 0$). Therefore we find, as expected, an extensive free energy, i.e., a free energy per site

$$f = \lim_{N\to\infty} \frac{F}{N} = -kT \log \lambda_1. \tag{4.15}$$

Exercise

Compute the partition function for the open chain, and verify that it leads to the same free energy per site f.

Now that we have computed exactly the free energy in the presence of an external field, we can easily obtain the magnetization

$$m = \langle \sigma_i \rangle = \frac{1}{N}\left\langle \sum_1^N \sigma_i \right\rangle = -\frac{1}{N}\frac{\partial F}{\partial h} \tag{4.16}$$

and let N go to infinity

$$m = -\frac{\partial f}{\partial h} = kT \frac{\partial \log \lambda_1}{\partial h}. \tag{4.17}$$

We can now let h go to zero and determine whether a spontaneous magnetization may exist for this one-dimensional model; expanding λ_1 for h small,

$$\lambda_1 = 2\cosh\beta J\left(1 + \frac{\beta^2 h^2}{2}\mathrm{e}^{2\beta J} + O(h^4)\right), \tag{4.18}$$

one concludes immediately that, for any temperature,

$$\lim_{h\to 0} m = 0. \tag{4.19}$$

The one-dimensional Ising model can be solved exactly, but it does not exhibit any broken symmetry. The system is paramagnetic at all temperatures, i.e., the magnetization vanishes with the applied field.

Exercises

1. *Find the transfer matrix for a one-dimensional Ising model, in which the spins may take the three values* $(+1, 0, -1)$.
2. *Find the transfer matrix for a one-dimensional Ising model with first and second neighbour interactions:*

$$E(\mathcal{C}) = -J\sum \sigma_i\sigma_{i+1} - K\sum \sigma_i\sigma_{i+2} - h\sum \sigma_i$$

   $(\sigma = \pm 1)$.
   Hint: consider the transfer of a pair to the neighbouring pair.
3. *Find the transfer matrix for a 'ladder' consisting of two parallel chains, i.e., call* $\sigma_i = \pm 1$ *the spins of the first chain,* $\tau_i = \pm 1$ *the spins on the second chain:*

$$E(\mathcal{C}) = -J\sum \sigma_i\sigma_{i+1} - J\sum \tau_i\tau_{i+1} - K\sum \sigma_i\tau_i.$$

4.2 One-dimensional Ising model: correlation functions

To understand better the absence of broken symmetry in this one-dimensional model, it is instructive to analyze the correlations between two spins located at sites i and $i+n$. Indeed, remember that the existence of a spontaneous magnetization could also manifest itself if, in zero field, one had

$$\lim_{n\to\infty} \langle \sigma_i\sigma_{i+n}\rangle \neq 0. \tag{4.20}$$

Let us compute

$$\langle \sigma_i\sigma_{i+n}\rangle = \frac{1}{Z}\sum_{\mathcal{C}} \sigma_i\sigma_{i+n}\mathrm{e}^{-\beta E(\mathcal{C})}. \tag{4.21}$$

The same identity (4.5) and the same transfer matrix technique leads to

$$\langle \sigma_i\sigma_{i+n}\rangle = \frac{1}{Z}\sum_{\sigma,\sigma'} \sigma T^n(\sigma,\sigma')\sigma' T^{N-n}. \tag{4.22}$$

Introducing the third Pauli matrix, the diagonal matrix

$$\tau = \begin{pmatrix} 1 & 0 \\ 0 & -1 \end{pmatrix}, \tag{4.23}$$

which is such that

$$[\tau \cdot A](\sigma, \sigma') = \sigma A(\sigma, \sigma'), \tag{4.24}$$

one finds

$$\langle \sigma_i \sigma_{i+n} \rangle = \frac{1}{Z} \mathrm{Tr}[\tau \mathcal{T}^n \tau \mathcal{T}^{N-n}]. \tag{4.25}$$

Let us introduce the eigenvectors of $\mathcal{T}$

$$\mathcal{T}|a\rangle = \lambda_a |a\rangle \quad \text{with } a = 1, 2 \tag{4.26}$$

then

$$\mathcal{T}^n = \lambda_1^n |1\rangle\langle 1| + \lambda_2^n |2\rangle\langle 2|. \tag{4.27}$$

Taking the limit of large N, fixed n, one obtains

$$\begin{aligned} \langle \sigma_i \sigma_{i+n} \rangle &= \frac{1}{\lambda_1^n} \langle 1|\tau \mathcal{T}^n \tau|1\rangle \\ &= |\langle 1|\tau|1\rangle|^2 + \left(\frac{\lambda_2}{\lambda_1}\right)^n |\langle 1|\tau|2\rangle|^2. \end{aligned} \tag{4.28}$$

For $h = 0$, the vector $|1\rangle$ is simply $\frac{1}{\sqrt{2}}(1, 1)$ and $\langle 1|\tau|1\rangle = 0$, whereas $\langle 1|\tau|2\rangle = 1$. Therefore, the correlation function decreases exponentially with the distance $r = na$ between the two sites

$$\langle \sigma_i \sigma_{i+n} \rangle = \mathrm{e}^{-r/\xi} \tag{4.29}$$

with a characteristic correlation length

$$\xi = a \left[\log \frac{\lambda_1}{\lambda_2} \right]^{-1}. \tag{4.30}$$

Therefore, there is a possibility of having a spontaneous magnetization, i.e., a non-vanishing limit of the correlation function when the two sites are far apart, only if the correlation length diverges, i.e., only if the two eigenvalues are degenerate. We have seen that this does not occur in this model, since λ_1 is strictly superior to λ_2 at any temperature. We shall see that this is a general characteristic of one-dimensional models with short-range interactions.

Exercise

For this one-dimensional chain with $h \neq 0$, show that

$$\lim_{N\to\infty} \langle \sigma_i \rangle = \langle 1|\tau|1\rangle$$

and that the result coincides with the magnetization given by $m = -\frac{\partial f}{\partial h}$.

4.3 Absence of phase transition in one dimension

Our conclusion of the previous section relied on an explicit exact solution. Is it conceivable that, within the same class of one-dimensional Z_2-invariant Hamiltonians, there might be other cases in which the symmetry could be spontaneously broken? Do we have to repeat the calculation for a system in which the spin takes three values $(+1, 0, -1)$ instead of two? Or if the first and second neighbours interact? Or if we have a ladder of two chains of spins? In all these cases one can define a transfer matrix (3×3 or 4×4 in the above examples) with real positive entries. (The matrix is not always real symmetric, and it may have complex eigenvalues.) Repeating the above analysis for a $k \times k$ transfer matrix (with k finite, independent of N), one finds again an exponential fall-off of the correlation function with distance with a correlation length given by (4.30), or rather $\xi = a\left[\log \frac{|\lambda_1|}{|\lambda_2|}\right]^{-1}$, in which $|\lambda_1| \geq |\lambda_2|$ are the two largest eigenvalue. Is it possible that a degeneracy of the largest eigenvalue could lead to a long-range order?

This possibility is ruled out by the Perron–Frobenius theorem, which states that, 'A matrix with real positive entries posseses one real, non-degenerate, eigenvalue, which exceeds the modulus of all the other eigenvalues. The corresponding eigenvector has real components, all of the same sign.'

(In fact, it is a version of this theorem for operators rather than matrices that one uses when one asserts that the spatial ground-state wave function of a Hamiltonian is non-degenerate and that the corresponding wave function is nodeless.)

Note that it is essential to assume that the rank of this transfer matrix is fixed, independent of the number of sites N. If we took a two-dimensional lattice $N = M^2$ and started to rank the sites as a chain going through the lattice, in which the lattice point (i, j) would be followed by $(i+1, j)$, but (M, j) by $(1, j+1)$, then the nearest-neighbour interactions on the two-dimensional lattice would couple sites at distance $M = \sqrt{N}$ and the transfer matrix would have a rank that goes to infinity with N. The Perron–Frobenius theorem still applies but nothing prevents the ratio $\frac{\lambda_2}{\lambda_1}$, which is now size dependent, to approach one in the large N limit.

Therefore, we can only state that for one-dimensional systems with short-range interactions there is no possibility of spontaneously broken symmetries. The argument would be parallel for a discrete symmetry group larger than Z_2.

4.4 A glance at the two-dimensional Ising model

Consider a portion of a square lattice of $N = L \times M$ sites. We can consider two adjacent rows of this lattice, each with M sites, and label the spins at the nodes of these rows $(\sigma_1, \sigma_2, \ldots, \sigma_M)$ for the first row and $(\sigma'_1, \sigma'_2, \ldots, \sigma'_M)$ for the next one. We define the matrix

$$\begin{aligned}\mathcal{T}(\{\sigma\}, \{\sigma'\}) &= \mathrm{e}^{\beta J(\sigma_1\sigma'_1+\sigma_2\sigma'_2+\cdots+\sigma_M\sigma'_M)} \\ &\times \mathrm{e}^{\frac{\beta J}{2}(\sigma_1\sigma_2+\sigma_2\sigma'_3+\cdots+\sigma_M\sigma_{M+1})} \\ &\times \mathrm{e}^{\frac{\beta J}{2}(\sigma'_1\sigma'_2+\sigma'_2\sigma'_3+\cdots+\sigma'_M\sigma'_{M+1})} \\ &\times \mathrm{e}^{\frac{\beta h}{2}(\sigma_1+\sigma'_1+\cdots+\sigma_M\sigma'_M)},\end{aligned} \tag{4.31}$$

in which we have assumed periodic boundary conditions $\sigma_{M+1} = \sigma_1$. It is straightforward to verify that the total Boltzmann weight of the lattice is

$$\mathrm{e}^{-\beta E(\mathcal{C})} = \mathcal{T}(\{\sigma_1\}, \{\sigma_2\})\mathcal{T}(\{\sigma_2\}, \{\sigma_3\}) \ldots \mathcal{T}(\{\sigma_L\}, \{\sigma_{L+1}\}), \tag{4.32}$$

assuming again p.b.c. in the vertical direction.

The partition function is then 'simply'

$$Z = \mathrm{Tr}\mathcal{T}^{\mathrm{L}} \tag{4.33}$$

but note that the matrix $\mathcal{T}$ is $2^M \times 2^M$, and that M and L go to infinity as $\sqrt{N}$. Finding the largest eigenvalue of $\mathcal{T}$ is thus a formidable problem and in fact it has never been done for the matrix (4.31), i.e., for a non-zero external field h. The solution of the problem by Onsager (1944), in the case of a zero field is a landmark in the history of physics. Indeed, Onsager's exact zero-field solution for the free energy exhibits a phase transition from a paramagnetic to a ferromagnetic phase. A spontaneous magnetization appears below some temperature T_c; the free energy is continuous but singular at T_c. We shall review below some of the results that came out of this solution. The original derivation of Onsager is not the easiest to read; the mapping to a problem of free fermions[1] gave rise later to much simpler solutions, and in modern days the two-dimensional Ising model is one of the simplest examples of conformal field theory.

4.5 Proof of broken symmetry in two dimensions (and more)

Before giving Peierls' proof,[2] one should start with an intuitive understanding of what is going on. Instead of starting with a vanishingly small external field, one can equivalently assume that the spins on the edge of the system are frozen, for

[1] D. Schultz, D. C. Mattis and E. H. Lieb, *Rev. Mod. Phys.*, **36** (1964) 856.

[2] R. Peierls, *Proc. Cambridge Phil. Soc.*, **32** (1936) 477.

instance, we assume that they are all pointing up (this is equivalent to a magnetic energy of order $\sqrt{N}$, i.e., to an external field of order $1/\sqrt{N}$). At zero temperature, the only contribution to the partition function comes from a configuration in which all spins point up. At higher temperatures, some bubbles of down spins will appear. A bubble of linear dimension L has a frontier with about L^{d-1} spins on its edge, in a lattice of dimension d. The interaction energy between two up spins being the same as for two down spins, the energy 'cost' of making such a bubble is of order $2JL^{d-1}$. The entropy for a bubble of this size is the logarithm of the number of ways of creating it, i.e., the number of closed paths of length L, whose logarithm grows as a constant times L. In one dimension, the entropy wins compared with the energy at any non-zero temperature. Therefore, bubbles are favoured and the ordered zero-temperature state is destroyed for any positive temperature. In dimensions greater than one, the energy cost wins compared to the entropy and the system will remain ordered up to some finite temperature.

I shall now give a real proof of the existencc of a spontaneous magnetization in the two-dimensional case below some finite temperature (following Ginibre's presentation of Peierls' proof). Again, we consider configurations in which the spins are frozen and $\uparrow$ on the boundary. A configuration $\mathcal{C}$ is made of islands of $\downarrow$ spins immersed in a sea of $\uparrow$ spins. These islands may be defined by closed polygons (Fig. 4.1) that enclose the $\downarrow$ spins. In fact, there is a one-to-one correspondence between a configuration $\mathcal{C}$ and a set of polygons: indeed, starting from the boundary with $\uparrow$ spins, one returns to them each time one crosses the border of a polygon. Thus, $\mathcal{C}$ is characterized by a set of disjunct connected polygons characterized by their location and their shape.

We can classify the polygons as follows: a polygon of perimeter b may be drawn at many different locations on the lattice, and may present different shapes.

We order using an index j, $1 \le j \le \nu(b)$, the various polygons $\mathcal{P}_b^{(j)}$ of perimeter b that one can draw on the lattice; $\nu(b)$ is the number of distinct polygons of

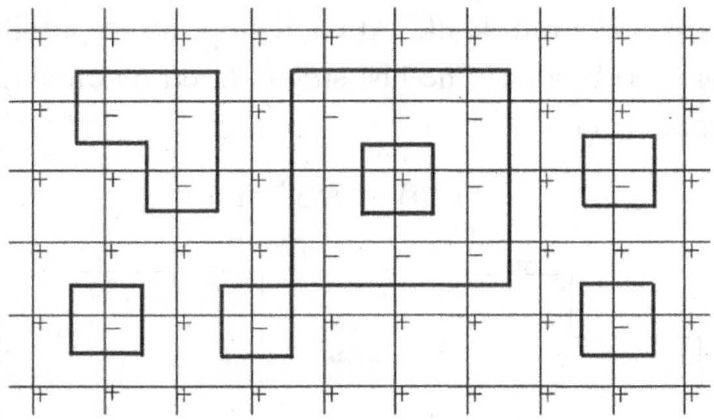

Figure 4.1 The polygons around the down spins

perimeter b that one can draw on the lattice. A polygon $\mathcal{P}_b^{(j)}$ encloses $N_b^{(j)}$ sites. A configuration $\mathcal{C}$ is thus a set of polygons

$$\mathcal{C} = \left(\mathcal{P}_{b_1}^{(j_1)}, \mathcal{P}_{b_2}^{(j_2)}, \ldots\right). \tag{4.34}$$

Let $N_-(\mathcal{C})$ and $N_+(\mathcal{C})$ be the number of down spins and up spins of $\mathcal{C}$. The problem consists of computing the average of $N(\mathcal{C})$ with the Boltzmann weight $\frac{1}{Z}\exp - \beta E(\mathcal{C})$. Indeed, the magnetization per site m is nothing but

$$m = \frac{1}{N}(\langle N_+(\mathcal{C}) - N_-(\mathcal{C})\rangle) = 1 - 2\frac{\langle N_-(\mathcal{C})\rangle}{N}. \tag{4.35}$$

The proof of spontaneous symmetry breaking (s.s.b.) consists in showing that, at sufficiently low temperatures, $\frac{\langle N_-(\mathcal{C})\rangle}{N} < \frac{1}{2}$.

It is convenient to introduce the characteristic function

$$\chi_b^{(j)}(\mathcal{C}) = \begin{cases} 1 \text{ if } \mathcal{P}_b^{(j)} \in \mathcal{C} \\ 0 \text{ if } \mathcal{P}_b^{(j)} \notin \mathcal{C}. \end{cases} \tag{4.36}$$

For a given configuration $\mathcal{C}$ one has

$$N_-(\mathcal{C}) \leq \sum_b \sum_{1\leq j\leq \nu(b)} \chi_b^{(j)}(\mathcal{C}) N_b^{(j)} \tag{4.37}$$

(the inequality corresponds to the cases in which one polygon encloses other polygons in its interior; in such cases, on the r.h.s. we are counting some up spins as down: hence the inequality rather than equality).

A simple bound on $N_b^{(j)}$ follows trivially from 'erasing' the angles, changing the polygon into a rectangle of the same perimeter but larger area (Fig. 4.2).

Since the rectangle of perimeter b with the largest area is a square, one obtains the simple bound

$$N_b^{(j)} \leq \left(\frac{b}{4}\right)^2. \tag{4.38}$$

Furthermore one can bound $\nu(b)$ as follows: starting at one of the N points of the lattice, one performs a random walk. At each step, three possibilities are allowed since backtracking is forbidden. The last step is fixed by closing the polygon. One then obtains the upper bound

$$\nu(b) \leq N 3^{b-1}. \tag{4.39}$$

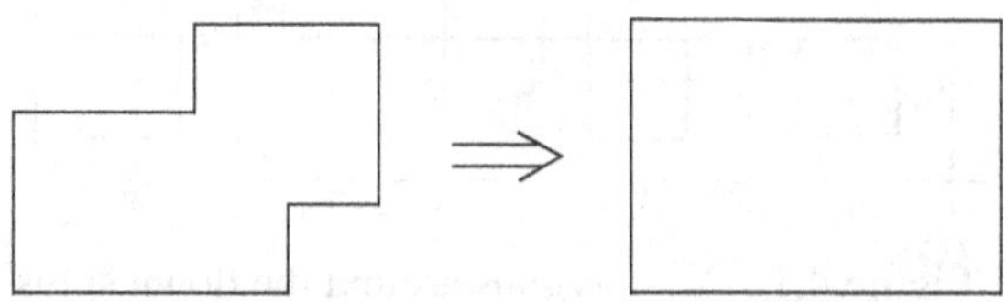

Figure 4.2 Replacement by a rectangle of equal perimeter and larger area

Finally one can bound $\langle \chi_b^{(j)}(\mathcal{C})\rangle$ as follows: by definition

$$\langle \chi_b^{(j)}(\mathcal{C})\rangle = \frac{\sum_{\mathcal{C}\supset\mathcal{P}_b^{(j)}} \mathrm{e}^{-\beta E(\mathcal{C})}}{\sum_{\mathcal{C}} \mathrm{e}^{-\beta E(\mathcal{C})}}. \tag{4.40}$$

Let us consider one of the configurations $\mathcal{C}$ present in the numerator of (4.40), i.e., containing among its polygons $\mathcal{P}_b^{(j)}$. To this configuration one may associate the configuration $\hat{\mathcal{C}}$, in which one has changed into their opposites all the spins which are inside $\mathcal{P}_b^{(j)}$. In $\hat{\mathcal{C}}$, the spins interact in the same way as they do in $\mathcal{C}$, except that in $\hat{\mathcal{C}}$ the polygon $\mathcal{P}_b^{(j)}$ has been erased. Therefore, we end up with the main relation

$$E(\hat{\mathcal{C}}) = E(\mathcal{C}) - 2bJ. \tag{4.41}$$

Note that the correspondence between $\mathcal{C}$ and $\hat{\mathcal{C}}$ is one-to-one for a given $\mathcal{P}_b^{(j)}$. One obtains an upper bound to the r.h.s. of (4.40) by limiting the configurations in the denominator to the $\hat{\mathcal{C}}$ configurations deduced from the $\mathcal{C}$ that intervene in the numerator. Thus, using (4.41)

$$\langle \chi_b^{(j)}(\mathcal{C})\rangle \le \frac{\sum_{\mathcal{C}\supset\mathcal{P}_b^{(j)}} \mathrm{e}^{-\beta E(\mathcal{C})}}{\sum_{\hat{\mathcal{C}}} \mathrm{e}^{-\beta E(\hat{\mathcal{C}})}} = \mathrm{e}^{-\beta 2bJ}. \tag{4.42}$$

We now put together all the inequalities:

$$\begin{aligned}
\langle N_-(\mathcal{C})\rangle &\le \sum_b \sum_{1\le j\le \nu(b)} \chi_b^{(j)}(\mathcal{C}) N_b^{(j)} \\
&\le \sum_b \mathrm{e}^{-\beta 2bJ} \sum_{1\le j\le \nu(b)} N_b^{(j)} \\
&\le \sum_b \mathrm{e}^{-\beta 2bJ} \left(\frac{b}{4}\right)^2 \sum_{1\le j\le \nu(b)} 1 \\
&\le \sum_b \mathrm{e}^{-\beta 2bJ} \left(\frac{b}{4}\right)^2 N 3^{b-1}.
\end{aligned} \tag{4.43}$$

The sum over b runs from $b = 4, 6, \ldots$ up to the largest polygon, for which b is about $4N$. Since we are looking for an upper bound, we can sum over b up to infinity, provided that the series converges. It does converge when

$$3\mathrm{e}^{-2\beta J} < 1 \tag{4.44}$$

and we assume that the temperature is such that this is true. The last sum of (4.43) is a monotonically decreasing function of β, which vanishes when $\beta \to \infty$. If we define a temperature $T_0 = \frac{1}{k\beta_0}$, such that

$$\sum_{b=4,6,\dots} e^{-2\beta_0 bJ} \left(\frac{b}{4}\right)^2 3^{b-1} = \frac{1}{2}, \tag{4.45}$$

we have shown that for $T < T_0$

$$\frac{\langle N_-(\mathcal{C})\rangle}{N} < \frac{1}{2}, \tag{4.46}$$

i.e., there is a spontaneous symmetry breaking leading to a positive magnetization m.

To generate a pure state, and not a mixture, we have broken infinitesimally the symmetry by freezing positively the spins on the boundary. One can prove that the spontaneous magnetization is the same if we had chosen to break the symmetry by a vanishingly small external field.

4.6 Correlation inequalities

Peierls' proof was relative to a specific two-dimensional lattice, nearest-neighbour interactions and spins equal to ± 1. It would be tedious to extend the proof to other Z_2-invariant Hamiltonians, such as interactions between first and second neighbours, three-dimensional lattices, etc. Fortunately some simple and general GKS inequalities (obtained by Kelly and Sherman, and extended by Griffiths) enable us to extend Peierls' result to other situations. Given a set of lattices

$$A = \{a_1, a_2, \dots, a_n\},$$

one denotes

$$\sigma_A = \sigma_{a_1} \cdot \sigma_{a_2} \cdots \sigma_{a_n} \tag{4.47}$$

and we take for the energy of a configuration the very general function

$$E(\mathcal{C}) = -\sum_A J_A \sigma_A, \tag{4.48}$$

in which we assume that all J_A are positive or zero. This is a compact way of writing this energy: it may contain a magnetic field for A reduced to a single site, first and second neighbour interactions, multispin interactions, etc. The dimension of the lattice is arbitrary.

We now state the following inequalities, whose derivation is left for an appendix:

- Inequality on expectation values

$$\langle \sigma_A \rangle \geq 0, \tag{4.49}$$

- Correlation inequality

$$\langle \sigma_A \sigma_B \rangle - \langle \sigma_A \rangle \langle \sigma_B \rangle \geq 0, \tag{4.50}$$

- Equivalent form of the correlation inequality (4.50): a simple differentiation of $\langle\sigma_A\rangle = 1/Z \sum \sigma_A \exp -\beta E(\mathcal{C})$ with respect to J_B gives

$$\frac{\partial\langle\sigma_A\rangle}{\partial J_B} = \beta(\langle\sigma_A\sigma_B\rangle - \langle\sigma_A\rangle\langle\sigma_B\rangle) \geq 0. \tag{4.51}$$

Let us look at a few consequences. If a Hamiltonian H_2 results from H_1 by the addition of terms of the form $\sum \delta J_A \sigma_A$, in which the δJ_A are positive, the magnetizations satisfy $m_2 \geq m_1$. From this, it follows that

1. The magnetization increases with the applied magnetic field.
2. The magnetization increases with the temperature decreases (an infinitesimal decrease of temperature $-\delta T$ is the addition to the Hamiltonian of $k\beta^2\delta T \sum J_A\sigma_A$).
3. The addition of attractive interactions between second neighbours implies that there is certainly a spontaneous magnetization at all temperatures at which there was one with only nearest-neighbour interactions.
4. At any temperature $T \leq T_c$ $(d=2)$ at which the two-dimensional Ising model possesses a spontaneous magnetization, the three-dimensional model with the same interactions also has a spontaneous magnetization, i.e.,

$$T_c(d=3) \geq T_c(d=2), \tag{4.52}$$

since one can consider the three-dimensional model as a stack of two-dimensional planes coupled by positive links that increase the magnetization.

In conclusion, we have seen that the ordinary principles of statistical mechanics do contain the possibility of spontaneous symmetry breaking, phase transitions, and their associated singularities without the need of any new assumption. The thermodynamic limit is responsible for the occurrence of such singular behaviour. This was a historical landmark, since it established that the usual Boltzmann–Gibbs principles of statistical mechanics allowed for phase transitions in the thermodynamic limit. We have seen that, at temperatures low enough, a spontaneous magnetization occurs; this spontaneous magnetization decreases with temperature. At high temperature, one can prove that this spontaneous magnetization vanishes identically. However, we are still left with two possibilities.

- The magnetization jumps from a finite value to zero at some temperature: a *first-order* transition (Fig. 4.3),
- The magnetization vanishes without discontinuity at the *critical temperature* T_c, a *second-order transition* (Fig. 4.3), which would be better named a continuous transition.

In two dimensions, the exact solution of Onsager, demonstrates a critical point. In three dimensions, numerous indications, experimental, numerical and theoretical (examined later) point to a continuous transition beyond any doubt, but there is no rigorous proof. In two dimensions, the exact solution allows one to determine the

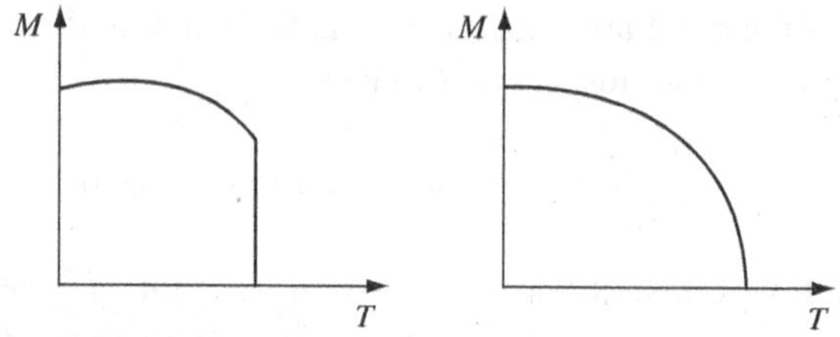

Figure 4.3 First-order and second-order transitions

exact critical behaviour. For instance, the spontaneous magnetization vanishes near T_c as $(T_c - T)^{1/8}$. The equivalent in three dimensions is close to a power of $\frac{1}{3}$. The main problem with exact solutions is that they are specific to models and they do not give any clue to the *universality* of the critical behaviour, i.e., the fact that this behaviour is the same for a whole class of Hamiltonians, a problem to which we shall address ourselves later in detail. It will then be necessary to leave these rigorous mathematics and use a different strategy.

4.7 Lower critical dimension: heuristic approach

I have given rigorous proof that for a discrete symmetry group, such as Z_2 of the Ising model, there was a spontaneous symmetry breaking in two dimensions but not in one. The lower critical dimension is thus equal to one. It has been known for several decades that the spontaneous breaking of a symmetry does not occur in low dimensions. Landau and Lifshitz[3] proved that it does not occur in one dimension (with short-range forces). Mermin and Wagner,[4] for statistical physics, and S. Coleman[5] for quantum field theory, showed that a continuous symmetry cannot be broken in two dimensions. Therefore, there is a lower critical dimension d_l, defined by the fact that above d_l, spontaneous symmetry breaking may take place for some non-vanishing range of temperature $0 < T < T_c$, i.e.,

$$T_c > 0 \qquad \text{for} \qquad d > d_l. \tag{4.53}$$

Let us compare the breaking of discrete versus continuous symmetries in determining d_l. In both cases, we begin at zero temperature with boundary conditions, such as the 'order parameter', fixed to a given value all over the boundary. Then we go to a slightly positive temperature, producing some excitations. The estimation

[3] L. D. Landau and E. M. Lifshitz, *Course of Theoretical Physics: Volume 5*, Statistical Physics, (Oxford: Pergamon Press, 1980).

[4] N. Mermin and H. Wagner, *Phys. Rev. Lett.*, **17** (1966) 1133.

[5] S. Coleman, *Comm. Math. Phys.*, **31** (1973) 25.

of the energy and the entropy of those excitations will provide a simple criterion for symmetry breaking:

- If the cost in energy is lower than the entropy of those excitations multiplied by the temperature, then even for very low T, the system will lower its free energy ($\Delta E - TS$) by destroying the zero-temperature ordered state. In such cases, T_c vanishes.
- If, however, the cost in energy is higher, at low enough temperatures, than the 'benefit' drawn from TS, the system will 'resist' the formation of disordering configurations and T_c will be finite: the dimension is then greater than d_l.

Let us apply this criterion successively to discrete and continuum symmetries.

Discrete symmetries

Let us consider the Ising model:

$$H = -J \sum_{\langle i,j \rangle} (\sigma_i \sigma_j - 1) \qquad \sigma_i = \pm 1, \tag{4.54}$$

in which the symbol $\langle i, j \rangle$ restricts the sum to nearest neighbours on a d-dimensional regular lattice.

Assume that the spins are frozen on the boundary, for instance to $\sigma_i = +1$ if $i \in$ boundary. Thus at $T = 0$ all the spins are up in the ground state. Let us estimate the cost in free energy to create a bubble of down spins (Fig. 4.4).

- If $d = 1$

  Consider a bubble of negative spins of size l, $l/L = O(1)$, so that the mean spin per site is now smaller than one. The energy and entropy associated with this bubble are:

$$\Delta E = 4J,$$
$$S \sim \log L,$$

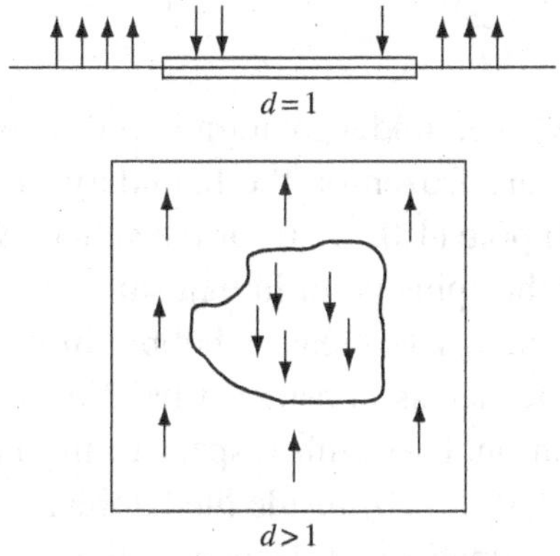

Figure 4.4 Bubbles of down spins

since there are $O(L)$ ways of choosing the location of this bubble. Therefore, the cost in free energy is:

$$\Delta F \sim 4J - T \log L. \tag{4.55}$$

For all non-zero temperatures, $\Delta F < 0$: the system lowers its free energy by making bubbles of opposite spins, and thus does not remain ordered.

- If $d = 2$

We want to estimate the energy and entropy associated with a bubble of linear size l. The excitation energy is simply equal to $2J$ multiplied by the perimeter of the bubble (which is of order l):

$$\Delta E \simeq 2Jl. \tag{4.56}$$

The entropy is estimated as follows: we fix an origin at one point of the perimeter of the bubble. There are about L^2 locations for that point. Then we perform a random walk, which starts and ends at this origin; the number of steps of this walk is simply the perimeter of the bubble. The number of such walks is about c^l if each point has c neighbours on this lattice. (This factor is much greater than the previous L^2.) The entropy is then proportional to $l \log c$. Therefore, the free energy estimate for these excitations is:

$$\Delta F \sim Jl - ATl \tag{4.57}$$

with some constant. At low enough temperature, $\Delta F > 0$, i.e., the entropy is not sufficient to overcome the excitation energy, and the system remains ordered. However, the entropy wins at higher temperatures.

Continuous symmetries

Let us consider, for definiteness, the Hamiltonian for n components with 'classical' spins,

$$H = -J \sum_{\langle i,j \rangle} (\vec{\sigma}_i \vec{\sigma}_j - 1) \qquad \vec{\sigma}^2 = \sum_{i=1}^{n} \sigma_i^2 = 1; \tag{4.58}$$

H is invariant under $O(n)$, i.e., under rotation in n dimensions.

Assume that the spins are frozen on the boundary of the sample, for instance, they all point to the north pole (1,0, ...,0) of the sphere S^{n-1}. At zero temperature H is minimum when all the spins point in that same direction. Again, we want to estimate the cost in free energy to create a bubble in which the spins correspond to a different ground state; let us consider a bubble in which the spins point in a direction that makes an angle θ with respect to the previous ground state, for instance $\vec{\sigma} = (\cos\theta, \sin\theta, 0, \ldots, 0)$ inside the bubble.

The entropy estimate is unchanged. However, the minimum cost in energy for creating this bubble is quite different.

(i) For the Ising model, it relies on the interfacial energy between the two ground states. The minimum energy between the ↑ and ↓ ground states is realized by a single plane of interface; the interfacial energy is thus proportional to L^{d-1}.

(ii) For the vector model, a hard wall between the two pure states, at which full rotation of θ would occur, would cost an energy $J(1 - \cos\theta)$ multiplied by L^{d-1} the number of spins in this wall. However, we can reach a much lower energy if, instead of a wall, one rotates progressively of θ/L in L successive steps. Indeed, for two nearby spins, this rotation costs an energy $J(1 - \cos\theta/L)$; in a 'plane' there are L^{d-1} sites. This is repeated L times to obtain the final rotation of θ. Therefore,

$$\Delta E = J(1 - \cos(\theta/L)) \qquad L^{d-1} \qquad L \sim J\theta^2 L^{d-2} \tag{4.59}$$

is proportional to L^{d-2} instead of L^{d-1}. The entropy estimate for this bubble is the same as for the Ising spins. Hence, for $d = 2$ $\Delta F < 0$ for any positive T: the system is disordered at any T. For $d > 2$, $\Delta E > T\Delta S$ for T small enough, i.e., there is an ordered state.

Again, this hand-waving argument is not a proof but the absence of long-range order follows rigorously from an inequality derived by Mermin and Wagner. In field theory, the impossibility of handling two-dimensional massless bosons, which would accompany continuous broken symmetries (see Chapter 13), has been demonstrated by Coleman.

The configurations that destroy the order (here the spin waves) are indeed massless Goldstone boson modes, and will be discussed in Chapter 13.

A word of warning: the absence of long-range order in two dimensions for a continuum symmetry does not imply that the free energy is analytic for all temperatures. For the two-dimensional XY model, in which each spin is confined to the circle $S_x^2 + S_y^2 = 1$, Kosterlitz and Thouless[6] proved that there is still a change of behaviour at a temperature T_c: for $T > T_c$ the spin–spin correlation function falls off exponentially with the distance; below T_c the correlation falls off as a power of the distance; it has to vanish at long distance, since there is no long-range order, but it vanishes slowly. Remarkably enough, the free energy is given by two functions, one above T_c, another one below. These functions, and all their derivatives, are equal at T_c.

4.8 Digression: Feynman path integrals, the transfer matrix and the Schrödinger equation

R. Feynman has introduced a revolutionary formulation of quantum mechanics based on path integrals. It turned out to be an invaluable tool for quantum field

[6] J. M. Kosterlitz and D. J. Thouless, Two-dimensional physics, *J. Phys. C: Solid State Phys.*, **6** (1973) L97.

theory, but let us take a brief look at the simple case of a particle in a potential, in dimension d. Then the probability amplitude for a particle located at point q_{in} at time 0 to be found at q_{f} at time τ is given by

$$A(q_{\text{in}}, q_{\text{f}}, \tau) = \sum_{\text{paths}} \mathrm{e}^{\mathrm{i}S[\text{path}]/\hbar}. \tag{4.60}$$

A path is a trajectory $q(t)$, i.e., any function that satisfies the boundary conditions $q(0) = q_{\text{in}}$ and $q(\tau) = q_{\text{f}}$; the action for this trajectory is

$$S[\text{path}] = \int_0^\tau \left[\frac{1}{2}m\dot{q}^2 - V(q)\right] \mathrm{d}t. \tag{4.61}$$

Clearly, the difficulty is to define properly the sum over all paths, an integral over all possible functions of t. The simplest way consists of discretizing the time interval τ into N steps,

$$\tau = N\epsilon, \tag{4.62}$$

and take as variable the positions

$$q_n = q(t = n\epsilon). \tag{4.63}$$

Then

$$\sum_{\text{paths}} \to \int \mathrm{d}q_1 \cdots \mathrm{d}q_{N-1}, \tag{4.64}$$

$$\int V(q)\mathrm{d}t \to \epsilon \sum_n V(q_n) \tag{4.65}$$

and

$$\int \dot{q}^2 \mathrm{d}t \to \epsilon \sum \left(\frac{q_{n+1} - q_n}{\epsilon}\right)^2. \tag{4.66}$$

Let us introduce the operator $\mathcal{T}$ defined by

$$\langle q'|\mathcal{T}|q\rangle = \mathrm{e}^{-\frac{\mathrm{i}\epsilon}{\hbar}V(q')} \mathrm{e}^{\frac{\mathrm{i}m}{2\epsilon\hbar}(q-q')^2}. \tag{4.67}$$

The discretized path integral involves a product of such $\mathcal{T}$ and the integrals over the intermediate q_n build up the Nth power of the operator:

$$A(q_{\text{in}}, q_{\text{f}}, \tau) = \lim_{\epsilon \to 0} \langle q_{\text{f}}|\mathcal{T}^{\tau/\epsilon}|q_{\text{i}}\rangle. \tag{4.68}$$

Therefore, like in the spin models, we have to find the eigenvalues of the transfer matrix $\mathcal{T}$. Let us note that, for a motion in dimension d,

$$\mathrm{e}^{\frac{\mathrm{i}m}{2\epsilon\hbar}(q-q')^2} = \langle q'|\mathrm{e}^{-\mathrm{i}\epsilon\frac{p^2}{2m\hbar}}|q\rangle, \tag{4.69}$$

in which p is the usual momentum operator, up to a multiplicative constant $\left(\left(\frac{m}{2\pi\hbar\epsilon}\right)^{d/2}e^{-i\pi d/4}\right)$ which has to be included in a definition of the measure for the integral over the intermediate q_n. Dropping this constant, we end up with the operator equation

$$\mathcal{T} = e^{-i\epsilon V(q)/\hbar}e^{-i\epsilon\frac{p^2}{2m\hbar}} \tag{4.70}$$

and, neglecting terms of order ϵ^2, we end up with

$$\mathcal{T} = e^{-i\epsilon H/\hbar}, \tag{4.71}$$

with the Hamiltonian $H = \frac{p^2}{2m} + V(q)$. The Schrödinger equation, which gives the eigenvalues of the Hamiltonian, is thus a means to diagonalize the transfer matrix for this path integral. This shows the usefulness of transfer matrices: here it allows one to connect the two formulations of quantum mechanics.

Appendix Derivation of the GKS inequalities

We follow here again the elegant derivation of Ginibre.[7]

Notation If A and B are two subsets of lattice points we denote as $A \ominus B$ the symmetric difference of the two sets consisting of the points in A not in B, and of the points of B not in A:

$$A \ominus B = (A \cap \bar{B}) \cup (B \cap \bar{A}), \tag{4.72}$$

in which $\bar{A}$ is the complementary set of A. The number of sites in A is denoted $|A|$. The product of the spins for all the sites belonging to A is σ_A. The value of σ_A for a configuration $\mathcal{C}$ is $\sigma_A(\mathcal{C})$. It is convenient, if A is the empty set to define $\sigma_\emptyset(\mathcal{C}) = 1$. The configuration $\mathcal{C}$ may be identified with the set C at which the spins point down.

Then one verifies easily that

$$\sigma_A(\mathcal{C}) = (-1)^{|A\cap C|} \tag{4.73}$$

and thus

$$\sigma_A(\mathcal{C}) = \sigma_C(A); \tag{4.74}$$

then

$$\begin{aligned}\sigma_A(\mathcal{C})\sigma_B(\mathcal{C}) &= \sigma_{A\ominus B}(\mathcal{C})\\ \sigma_C(\mathbf{A})\sigma_C(\mathbf{B}) &= \sigma_C(\mathbf{A}\ominus\mathbf{B})\end{aligned} \tag{4.75}$$

[7] J. Ginibre, *Phys. Rev. Lett.*, **23** (1969) 828.

and

$$\sum_{\mathcal{C}} \sigma_A(\mathcal{C}) = \begin{cases} 2^N \text{ if } A = \emptyset \\ 0 \text{ if } A \neq \emptyset. \end{cases} \tag{4.76}$$

Exercise

Verify this last result.

1. Let us come now to the first inequality:

$$\langle \sigma_B \rangle = \frac{1}{Z} \sum_{\mathcal{C}} \sigma_B(\mathcal{C}) e^{\beta \sum_A J_A \sigma_A(\mathcal{C})}. \tag{4.77}$$

Expanding the exponential, we obtain a sum of terms for which one needs to compute

$$\sum_{\mathcal{C}} \sigma_B(\mathcal{C}) \sigma_{A_1}(\mathcal{C}) \ldots \sigma_{A_n}(\mathcal{C}).$$

Applying, successively, (4.75) and (4.76) we see that this term is positive or zero.

2. The derivation of the second inequality is more subtle. We start with the definition

$$\begin{aligned} \langle \sigma_A \sigma_B \rangle - \langle \sigma_A \rangle \langle \sigma_B \rangle &= \frac{1}{Z} \sum_{\mathcal{C}} \sigma_A(\mathcal{C}) \sigma_B(\mathcal{C}) e^{\beta \sum_D J_D \sigma_D(\mathcal{C})} \\ &- \frac{1}{Z^2} \sum_{\mathcal{C},\mathcal{C}'} \sigma_A(\mathcal{C}) \sigma_B(\mathcal{C}') e^{\beta \sum_D J_D (\sigma_D(\mathcal{C}) + \sigma_D(\mathcal{C}'))}. \end{aligned} \tag{4.78}$$

We replace $1/Z$ in the first term with $1/Z^2 \sum_{\mathcal{C}'} \exp(\beta \sum_D J_D \sigma_D(\mathcal{C}'))$ and obtain

$$\begin{aligned} \langle \sigma_A \sigma_B \rangle - \langle \sigma_A \rangle \langle \sigma_B \rangle = \frac{1}{Z^2} \sum_{\mathcal{C},\mathcal{C}'} &[\sigma_{A \ominus B}(\mathcal{C}) - \sigma_A(\mathcal{C}) \sigma_B(\mathcal{C}')] \\ &\times e^{\beta \sum_D J_D (\sigma_D(\mathcal{C}) + \sigma_D(\mathcal{C}'))}. \end{aligned} \tag{4.79}$$

Define the set $\mathcal{C}''$

$$\mathcal{C}'' = \mathcal{C} \ominus \mathcal{C}', \tag{4.80}$$

which implies that

$$\mathcal{C}' = \mathcal{C} \ominus \mathcal{C}''. \tag{4.81}$$

Then

$$\begin{aligned} \sigma_{A \ominus B}(\mathcal{C}) - \sigma_A(\mathcal{C}) \sigma_B(\mathcal{C}') &= \sigma_{A \ominus B}(\mathcal{C}) - \sigma_A(\mathcal{C}) \sigma_B(\mathcal{C} \ominus \mathcal{C}'') \\ &= \sigma_{A \ominus B}(\mathcal{C}) - \sigma_A(\mathcal{C}) \sigma_B(\mathcal{C}) \sigma_B(\mathcal{C}'') \\ &= \sigma_{A \ominus B}(\mathcal{C})(1 - \sigma_B(\mathcal{C}'')). \end{aligned} \tag{4.82}$$

Furthermore,

$$\sigma_D(\mathcal{C}) + \sigma_D(\mathcal{C}') = \sigma_D(\mathcal{C}) + \sigma_D(\mathcal{C} \ominus \mathcal{C}''), \tag{4.83}$$

leading, finally, to

$$\langle\sigma_A\sigma_B\rangle - \langle\sigma_A\rangle\langle\sigma_B\rangle = \frac{1}{Z^2}\sum_{\mathcal{C}''}(1-\sigma_B(\mathcal{C}''))$$
$$\times \sum_{\mathcal{C}} \sigma_{A\ominus B}(\mathcal{C}) e^{\beta\sum_D J_D(1+\sigma_D(\mathcal{C}''))\sigma_D(\mathcal{C})}. \qquad (4.84)$$

The last expression may be read as follows: we are computing the expectation value of $\sigma_{A\ominus B}$ with a Hamiltonian in which the interactions J_D have been replaced by $\tilde{J}_D = J_D(1+\sigma_D(\mathcal{C}''))$, which are still non-negative. Therefore, from the first inequality, we learn that the sum over $\mathcal{C}$ produces a non-negative number. The final sum over $\mathcal{C}''$ with the positive weight $(1-\sigma_B(\mathcal{C}''))$ gives a non-negative result.

5

High-temperature and low-temperature expansions

These expansions have been used a great deal for getting numerical estimates of the critical indices. They also allow us to connect critical phenomena to other domains of science, such as the geometry of polymers or percolation problems: it is thus useful to study them in some detail.

5.1 High-temperature expansion for the Ising model

When a variable, such as σ, takes only two values, any function of σ is linear. Indeed

$$f(\sigma) = \frac{1}{2}\left[f(1) + f(-1)\right] + \frac{1}{2}\left[f(1) - f(-1)\right]\sigma \tag{5.1}$$

(if σ took n distinct values, any function of σ would be a polynomial of order $(n-1)$). In particular,

$$\begin{aligned} e^{\beta h\sigma} &= \cosh(\beta h) + \sinh(\beta h)\sigma = \cosh(\beta h)(1 + \tau\sigma) \\ e^{\beta J\sigma\sigma'} &= \cosh(\beta J) + \sinh(\beta J)\sigma\sigma' = \cosh(\beta J)(1 + w\sigma\sigma'), \end{aligned} \tag{5.2}$$

with

$$\tau = \tanh(\beta h) \qquad\qquad w = \tanh(\beta J). \tag{5.3}$$

At high temperature τ and w are small and it is much easier to expand in powers of these variables rather than in powers of β.

If we apply the identities (5.2) to the Boltzmann weight,

$$e^{\beta J\sum_{\langle i,j\rangle}\sigma_i\sigma_j + \beta h\sum_i \sigma_i},$$

we obtain the partition function:

$$Z = [\cosh(\beta J)]^L[\cosh(\beta h)]^N \sum_C \prod_{\langle i,j\rangle}(1 + w\sigma_i\sigma_j)\prod_i(1 + \tau\sigma_i), \tag{5.4}$$

in which L is the number of links between nearest neighbours. Let us note the following simple identity:

$$\sum_C \sigma_a \sigma_b \ldots \sigma_n = \begin{cases} 2^N & \text{if each index appears an even number of times} \\ 0 & \text{otherwise.} \end{cases} \tag{5.5}$$

An arbitrary term $\tau^l w^n$ of the expansion of (5.4) in powers of w and τ will thus give a non-zero contribution if and only if every site that it generates appears an even number of times. One can represent a non-zero term by putting a cross on the sites that gave a factor τ and a bar for the links that gave a w (Fig. 5.1). Note that a given site has at most one cross, and a given link at most one bar.

To make the discussion simpler, let us limit ourselves to $h = 0$, i.e., $\tau = 0$, and to two dimensions. Then the expansion is simply represented by a set of closed polygons on the lattice. The power of w, related to a set of polygons, is simply their total perimeter. Therefore, one finds

$$Z = [\cosh(\beta J)]^L 2^N \sum_p w^p g_N(p), \tag{5.6}$$

in which $g_N(p)$ is the number of closed polygons, connected or disconnected, that may be drawn on the lattice. By convention, we define $g_N(0) = 1$ and, obviously $g_N(2) = 0,\ \ g_N(2k+1) = 0$. For a square lattice one verifies immediately that $g_N(4) = N,\ g_N(6) = 2N$, since there are two types of polygon of perimeter equal to 6, the horizontal one and the vertical one. (I am here giving the leading term for N large; the subleading terms are of order $\sqrt{N}$.)

One has to go to order w^8 to discover the delicate nature of the problem. Indeed, there are both connected polygons of perimeter 8, and a set of two disconnected squares, which both contribute to $g_N(8)$. The number of pairs of squares is proportional to N^2, so $g_N(8)$ has the form

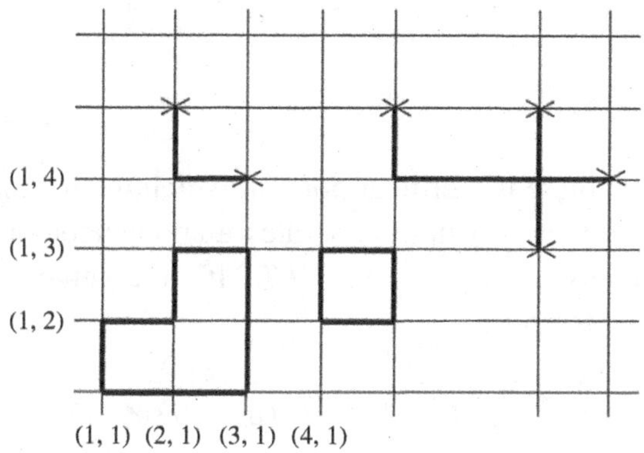

Figure 5.1 One term in the high-T expansion

$$g_N(8) = aN^2 + bN. \tag{5.7}$$

Now, we should remember that $\log Z$ is extensive and

$$\log Z = 2N \log\cosh\beta J + N\log 2 + \log\left(1 + \sum_{p\geq 4} g_N(p)w^p\right) \tag{5.8}$$

(for a square lattice $L = 2N$) and, expanding up to order 8,

$$\log Z = 2N\log\cosh\beta J + N\log 2 + g_N(4)w^4 + g_N(6)w^6 + \left[g_N(8) - \frac{1}{2}g_N(4)^2\right]w^8 + 0(w^{10}). \tag{5.9}$$

There, one sees that up to order 6 the extensivity of $\log Z$ is manifest, but at order 8 the coefficient is the difference of two terms of order N^2, and this N^2 part should cancel (which means that $a = 1/2$ in the above expression for $g_N(8)$, as expected, since there are about N^2 pairs of squares but each pair is computed twice), but the remaining term of order N is then quite difficult to compute. (A word of warning: $\log Z$ is extensive but, unfortunately, it is not true that its expansion is given simply in terms of connected polygons of perimeter p. This comes from the fact that in the expansion (5.8) a link cannot be occupied more than once.)

We shall not enter into the technical delicacies of these calculations. We simply note that one can obtain by a similar method the magnetization $m = \frac{1}{\beta N}\frac{\partial \log Z}{\partial h}$, which vanishes with h as long as the series in w converges, i.e., $T \geq T_c$. Similarly, one can obtain the *magnetic susceptibility* in zero field

$$\chi = \left.\frac{\partial m}{\partial h}\right|_{h=0} = \beta\left(1 + \sum_{n\geq 1} a_n w^n\right). \tag{5.10}$$

Elaborate work has resulted in the computation of about 20 coefficients a_n for the three-dimensional cubic lattice: $a_1 = 6, a_2 = 30, a_3 = 150, \ldots, a_{17} = 401\,225\,391\,222, \ldots$.

Exercise

Show that $a_1 = 6$.

If one assumes that there is a critical point at which the magnetic susceptibility diverges, since an infinitesimal field generates a spontaneous magnetization, the series (5.10) must diverge at $w_c = \tanh J/kT_c$. If we assume a critical behaviour in the vicinity of T_c of the form

$$\chi = \frac{\chi_0}{(T - T_c)^\gamma} = \frac{\chi_1}{(w_c - w)^\gamma}, \tag{5.11}$$

the series for

$$\frac{d\log\chi}{dw} = \sum b_n w^n \tag{5.12}$$

will present a pole at w_c with residue γ. There are standard numerical methods for extracting the singularity from an expansion. The knowledge of M coefficients a_n for (4.40), determines also M coefficients for (5.12) and one can replace the series by a ratio of two polynomials $P(w)/Q(w)$ of respective degrees m_1 and m_2, with $m_1 + m_2 = M$ (Padé approximant). The closest zero of Q will give an approximation of w_c and the residue at this zero an estimate of γ. This method has been used abundantly and has provided accurate estimates of exponents, such as γ; however, it does not produce any clue to the *universality* of the exponents, a fascinating fact of nature that we shall examine in detail later.

5.1.1 Continuous symmetry

A simple generalization of the model, of important physical relevance, consists of introducing at each lattice site i an n-component vector $\vec{S}_i$ of fixed length. For later convenience, we choose the normalization

$$(\vec{S}_i)^2 = n. \tag{5.13}$$

A configuration $\mathcal{C}$ now consists of N vectors of this S^{n-1} sphere. Again, we assume nearest-neighbour interactions and the energy of a configuration is

$$E(\mathcal{C}) = -J \sum_{\langle i,j \rangle} \vec{S}_i \cdot \vec{S}_j. \tag{5.14}$$

This interaction is invariant under the $O(n)$ group, which rotates the N 'classical spins' simultaneously. Note that the $n = 1$ model reduces to the Ising model. The partition function is obtained through an integration in the phase space of the N vectors $(\vec{S}_1, \ldots, \vec{S}_N)$ in which each vector varies with a uniform measure on the S^{n-1} sphere. This measure, $d\mu(\vec{S})$, is normalized to one,

$$\int d\mu(\vec{S}) = 1, \tag{5.15}$$

and one verifies easily that

$$\begin{aligned} &\int d\mu(\vec{S})\, S^\alpha S^\beta = \delta^{\alpha\beta} \\ &\int d\mu(\vec{S})\, S^\alpha S^\beta S^\gamma S^\delta = \frac{n}{n+2}(\delta^{\alpha\beta}\delta^{\gamma\delta} + \delta^{\alpha\gamma}\delta^{\beta\delta} + \delta^{\alpha\delta}\delta^{\beta\gamma}). \end{aligned} \tag{5.16}$$

These identities rely on rotational invariance, and the normalizations are easy to find (for instance, set $\alpha = \beta$ and sum over α from 1 to n). The high-temperature expansion involves now products of links $S_i^\alpha S_j^\alpha$ and since, for each site, each S^α must appear an even number of times, one again generates closed polygons with an internal index conserved all along the polygon: $S_i^\alpha S_j^\alpha S_j^\alpha S_k^\alpha \ldots S_l^\alpha S_i^\alpha$; however, a link may now be chosen several times.

Exercise

Compute the leading term of the correlation function $\langle \vec{S}(0)\vec{S}(\vec{r})\rangle$ for large T.

We shall not go further in this direction, but in the next chapter we shall study an important, special case of this problem.

5.2 Low-temperature expansion

We now start in the low-temperature phase of an Ising model. We assume that the boundary conditions have selected a pure state of positive magnetization. Let $E_0 = -JL$ be the lowest energy state in which all the spins are up (L is the number of links). When $T \to 0$, i.e., $\beta \to \infty$, the contribution in which all the spins point up is dominant, $Z \simeq \mathrm{e}^{-\beta E_0}$. The configurations of increasing energy correspond to one spin turned down, energy $E_0 + 8J$ in a two-dimensional square lattice, the two spins down, etc.

For an arbitrary configuration C one can separate the spins $\downarrow$ from the $\uparrow$ by a surface in three dimensions, a set of polygons in two dimensions that separates them from each other. In two dimensions, these polygons are drawn on the 'dual' lattice, i.e., the lattice made of the centres of the squares (Fig. 5.2).

If the set of polygons (we now stay in two dimensions) that separate the two species has a perimeter b, the configuration has an energy

$$E(C) = E_0 + 2bJ. \tag{5.17}$$

Therefore, we end up with the large β expansion

$$Z = \mathrm{e}^{\beta LJ} \sum_b \gamma_N(b)\mathrm{e}^{-2\beta bJ}, \tag{5.18}$$

in which $\gamma_N(b)$ is the number of sets of polygons with total perimeter b.

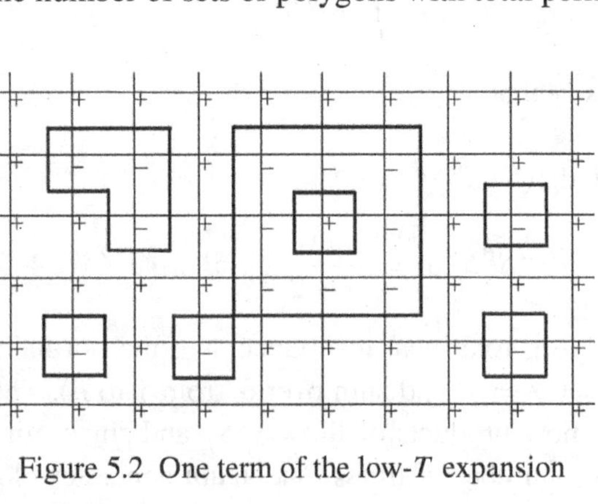

Figure 5.2 One term of the low-T expansion

5.2.1 Kramers–Wannier duality

The two-dimensional square lattice is identical to its dual, the lattice of the centres. If we put together the high-T expansion (5.6) and the low-T expansion (5.25), we see two expansions, one in terms of $w = \tanh(\beta J)$ and one in terms of $e^{-2\beta J}$ with the same coefficients $g_N(b)$. Therefore, if we associate to an inverse temperature β its 'dual' $\beta^\star$ defined as

$$\tanh \beta J = e^{-2\beta^\star J} \tag{5.19}$$

we find

$$\begin{aligned} Z(\beta)[\cosh \beta J]^{-2N} 2^{-N} &= \sum (\tanh \beta J)^p g_N(p) \\ &= \sum e^{-2p\beta^\star J} g_N(p) \\ &= Z(\beta^\star) e^{-2\beta^\star N J}. \end{aligned} \tag{5.20}$$

Note that the relation (5.19) is symmetric in the two variables, as becomes obvious when it is cast into the equivalent form

$$\sinh(2\beta J)\sinh(2\beta^\star J) = 1, \tag{5.21}$$

and thus $(\beta^\star)^\star = \beta$. From (5.20) we obtain for the free energy per site $f = 1/N \log Z$

$$f(\beta) - \log[2\cosh^2 \beta J] = f(\beta^\star) - 2\beta^\star J. \tag{5.22}$$

Exercise
Check that repeating the application of the duality relation (5.19) is consistent and useless.

If one assumes that the free energy is singular at the critical (inverse) temperature β_c, and that this singularity is unique, then one must have $\beta_c^\star = \beta_c$, which leads readily to the value

$$2\beta_c J = \log(1 + \sqrt{2}). \tag{5.23}$$

This duality, obtained by Kramers and Wannier, preceded by four years Onsager's exact and complete solution for the free energy, which confirmed the above value for the critical temperature.

Exercise
Start with an Ising model on a triangular planar lattice.

- *Show that the low-T expansion involves polygons on a hexagonal lattice.*
- *Show the identity*

$$e^{J(\sigma_1+\sigma_2+\sigma_3)} + e^{-J(\sigma_1+\sigma_2+\sigma_3)} = A e^{J'(\sigma_1\sigma_2+\sigma_2\sigma_3)+\sigma_3\sigma_1}$$

and determine A and J'.

- *Show that the previous identity allows one to sum over half the spins of the hexagonal lattice and return to a triangular lattice.*
- *Show that, after the previous summation, there is also a Kramers–Wannier duality for this triangular lattice and determine its critical temperature.*

5.3 Low-temperature expansion for a continuous symmetry group

Again, one can envisage all kinds of continuous symmetries, but we shall limit the discussion to the rotation group O_n in n dimensions. As seen in Section 4.7, symmetry breaking occurs only if the dimension of the lattice is greater than two. For definiteness we take again an n-component order parameter $\vec{S}$ of unit length:

$$\vec{S}^2 = 1. \tag{5.24}$$

A configuration $\mathcal{C}$ consists of a collection of N vectors $\vec{S}_i$ (N is the number of lattice sites) with an energy

$$\begin{aligned} E(\mathcal{C}) &= -\sum_{i,j} J(|\vec{i}-\vec{j}|)\vec{S}_i \cdot S_j \\ &= \frac{1}{2}\sum_{i,j} J(|\vec{i}-\vec{j}|)(\vec{S}_i - \vec{S}_j)^2 + \text{constant}, \end{aligned} \tag{5.25}$$

in which the interaction is attractive, short ranged and invariant by translation.

Assume that the boundary conditions select, at zero temperature, the pure state in which all the spins point along the first axis; namely they are all along the vector $(1, 0, \ldots, 0)$. At small non-zero temperature we parametrize each spin as

$$\vec{S} = (\sigma, \sqrt{T}\pi). \tag{5.26}$$

The constraint (5.24) allows us to eliminate σ:

$$\sigma = (1 - T\pi^2)^{1/2} = 1 - \frac{1}{2}T\pi^2 + O(T^2) \tag{5.27}$$

(we have neglected the 'large' fluctuations in which σ would become negative). One then finds

$$\begin{aligned} (\vec{S}_i - \vec{S}_j)^2 &= (\sigma_i - \sigma_j)^2 + T(\pi_i - \pi_j)^2 \\ &= T(\pi_i - \pi_j)^2 + \frac{1}{4}T^2(\pi_i^2 - \pi_j^2)^2 + O(T^3). \end{aligned} \tag{5.28}$$

One obtains thereby a systematic expansion of $E(\mathcal{C})/kT$ in powers of T, in which appear only the $(n-1)$ components of π. A priori, the components of these vectors vary in a bounded domain ($\pi^2 < 1/T$), but the error resulting from ignoring this constraint and integrating the components of π from $-\infty$ to $+\infty$ is of order $e^{-1/T}$ and it has no effect on the asymptotic expansion in powers of T. At lowest order

$$\left(\frac{E(\mathcal{C})}{T}\right)^{(0)} = \frac{1}{2}\sum_{i,j} J(|\vec{i}-\vec{j}|)(\pi_i - \pi_j)^2. \tag{5.29}$$

The measure of integration over the $(n-1)$ components of π is deduced from the uniform measure on the original sphere, namely $\delta(\sigma^2 + T\pi^2 - 1)$ which, after integration over σ yields the measure $d\pi(1 - T\pi^2)^{-1/2}$.

Exercise

Compute, up to first order in T, the correlation function $\langle \vec{S}(\vec{r})\vec{S}(0)\rangle$.

The formulation of the spin model (5.25), for a 'spin' confined to a given manifold, here the S^{n-1} sphere defined by the constraint $\vec{S}^2 = 1$, is named the *non-linear sigma model*. We shall return to it in Chapter 13.

6

Some geometric problems related to phase transitions

Some unexpected relationships between problems of completely different nature a priori follow from the models that we have considered in the previous chapters. We shall examine here two of geometric nature: *self-avoiding walks* and *percolation*. Again, the technique of high-temperature expansions introduced in the previous chapter will be a guide to establish these surprising connections.

6.1 Polymers and self-avoiding walks

Polymers are long molecules composed of N monomers, and N is here a large number. When the interaction between the monomers is completely negligible, the geometric configuration is similar to a *Brownian* chain, i.e., a simple chain of N successive independent steps, made at random. This is analogous to a random walk on a lattice. It is elementary to verify that the typical end-to-end distance between the endpoints of this chain is of order $N^{1/2}$.

Exercise

Prove the $N^{1/2}$ law for a Brownian chain. (Hint: write $\vec{R} = \vec{x}_1 + \vec{x}_2 + \cdots + \vec{x}_N$, in which the $\vec{x}_i$ are independent random unit vectors joining two successive points of the chain and compute $\langle \vec{R}^2 \rangle^{1/2}$.)

Another way of expressing this $\sqrt{N}$ law is through the notion of *fractal dimension* of the chain.[1] This notion may be simply related to a mass to length relationship: an object with uniform density of mass and typical diameter L has a fractal dimension d_f if the total mass M scales with L as

$$M \sim L^{d_f}. \tag{6.1}$$

[1] A concept introduced by the mathematician Haussdorf and whose ubiquity in the physical world was underlined by B. Mandelbrot.

This definition coincides with the usual notion of dimension for a line, area or volume, but it may be used in more complex geometric situations. For the Brownian chain, if each monomer has a unit mass, M is proportional to N, and the typical length scale L is of order of the end-to-end distance, which scales like $\sqrt{N}$, or gyration ratio R, if in polymer language. Therefore, this Brownian chain has a mass proportional to L^2, i.e., a fractal dimension equal to two: it is more like an area than a line.

When the monomers are interacting with each other, the problem is much more difficult. If the interaction is attractive, the chain collapses to a dense blob. If the interaction is repulsive, often a simple 'excluded volume' effect, the chain is more extended than the previous Brownian chain. The simplest model for this repulsion is to imagine a random walk on a lattice, in which it is forbidden to visit again a previously visited site: this is called a self-avoiding walk. It is a difficult problem because it is 'non-Markovian', i.e., the nth step depends on the whole past of the chain, and not simply of the last (or the last few) steps, as for the Brownian chain. It is expected that the gyration ratio R will be larger, as a function of N, because of this geometric repulsion. If one defines here too a critical exponent ν through the relation

$$R \sim N^{\nu}, \tag{6.2}$$

one expects a value of ν greater than $1/2$, and such a chain has a fractal dimension $d_{\mathrm{f}} = 1/\nu < 2$.

It has been shown by P. G. de Gennes[2] that this problem of the self-avoiding walk may be mapped onto a singular $n = 0$ limit of the n-vector model considered in the previous chapter. Let us return to Section 5.1.1 for n-vector models. It consists of 'spin' vectors of fixed length and random orientation on a lattice of N sites: we choose again the normalization

$$(\vec{S}_i)^2 = n. \tag{6.3}$$

A configuration $\mathcal{C}$ consists of a random vector at each lattice site and, assuming nearest-neighbour interactions, the energy of a configuration is

$$\beta E(\mathcal{C}) = -J \sum_{\langle i,j \rangle} \vec{S}_i \cdot \vec{S}_j. \tag{6.4}$$

We want to perform a high-temperature expansion for this model:

$$Z = \int \mathrm{d}\mu(\vec{S}_1) \cdots \mathrm{d}\mu(\vec{S}_N) \sum_0^{\infty} \frac{J^k}{k!} \left(\sum_{\langle i,j \rangle} \vec{S}_i \cdot \vec{S}_j \right)^k , \tag{6.5}$$

[2] P. G. de Gennes, *Phys. Lett. A*, **38** (1972) 339.

in which $\mathrm{d}\mu(\vec{S})$ is the uniform measure on the sphere (6.3), normalized to $\int \mathrm{d}\mu(\vec{S}) = 1$. Again, each term of the above expansion may be represented by a 'graph' on the lattice, which indicates how many times a given link has been chosen in the expansion, weighted by the expectation values over the spheres for the attached spins.

The integrals over the sphere for one single spin have a generating function

$$f_n(\vec{x}) = \int \mathrm{d}\mu(\vec{S})\, \mathrm{e}^{\vec{x}\cdot\vec{S}}. \tag{6.6}$$

The successive derivatives of $f_n(\vec{x})$ with respect to the components x^α at $\vec{x} = 0$ provide the expectation values for one spin. Rotational invariance implies that $f_n(\vec{x})$ is a function of the length x of the vector $\vec{x}$. Taking the vector $\vec{x}$ along one axis, the standard integration measure on the sphere provides the result

$$f_n(x) = \frac{\int_0^\pi \mathrm{d}\theta\, \sin^{n-2}\theta\, \mathrm{e}^{x\cos\theta}}{\int_0^\pi \mathrm{d}\theta\, \sin^{n-2}\theta}, \tag{6.7}$$

from which standard beta-function integrals yield the expansion

$$f_n(x) = 1 + \sum_1^\infty \frac{n^p}{2^p p! n(n+2)\dots(n+2p-2)} x^{2p}. \tag{6.8}$$

We are going to specialize this result to an a priori meaningless limit: *the $n = 0$ limit*. If we take this limit on the result (6.8), we find

$$f_0(x) = 1 + \frac{1}{2}x^2. \tag{6.9}$$

Exercise

Derive this result directly. Hint: relate the Laplacian of f_n to f_n for integer dimension n. Transform the relation to a differential equation and use the boundary conditions $f_n(0) = 1$, $f_n''(0) = 1$, in view of the first identity (5.16); take the $n = 0$ limit.

The result (6.9) implies that, since the degree of $f_0(x)$ is 2, the $n = 0$ limit of the above graph, in the high-temperature expansion, consists of connected or disconnected polygons, which do not pass more than twice through a given site. Consequently, the polygons generated by the high-T expansion of the partition function are non-intersecting. In addition, for any non-intersecting closed polygon the internal index of the spin is conserved around the polygon, since $\langle S_\alpha S_\beta\rangle$ vanishes for $\alpha \neq \beta$: therefore, the summation over a given internal index of the spin gives a factor n. This gives the trivial result

$$\lim_{n\to 0} Z(J) = 1. \tag{6.10}$$

However, let us go a bit further and compute the spin–spin correlation function for a given spin index:

$$G^{1,1}(r;J) = \langle S_0^1 S_r^1 \rangle = Z^{-1} \int \prod_i \mathrm{d}\mu(\vec{S}_i) S_0^1 S_r^1 \mathrm{e}^{J\sum_{\langle i,j\rangle} \vec{S}_i \cdot \vec{S}_j} \tag{6.11}$$

in the high-temperature expansion. The same analysis of the expansion and of the $n = 0$ limit leads to

$$\lim_{n\to 0} G^{1,1}(r;J) = \sum_p J^p c_p(r), \tag{6.12}$$

in which $c_p(r)$ is the number of paths joining the points 0 and r in p steps *without self-intersection*.

The relation (6.12) allows one to recover information on the asymptotics of self-avoiding walks. For instance, we can sum over $\vec{r}$:

$$c_p = \sum_{\vec{r}} c_p(\vec{r}) \tag{6.13}$$

is the total number of random walks of length p. If we sum the l.h.s. of (6.12) over $\vec{r}$ we obtain the response at site 0 of an applied uniform magnetic field, i.e., the magnetic susceptibility, which, for a magnet, diverges when J approaches the critical inverse temperature J_c, as

$$\chi \sim (J_\mathrm{c} - J)^{-\gamma}. \tag{6.14}$$

This singularity is related to the behaviour of c_p for large p, since the previous singularity implies that:

$$c_p \sim p^{\gamma-1} a^p, \tag{6.15}$$

in which $a = 1/J_\mathrm{c}$. Similarly, the average size of a self-avoiding walk consisting of p steps is given by

$$R_p = \left[\frac{1}{c_p}\sum_{\vec{r}} r^2 c_p(\vec{r})\right]^{1/2}. \tag{6.16}$$

If the magnetic correlation falls off exponentially with a correlation length ξ, which diverges near J_c as $(J_\mathrm{c} - J)^{-\nu}$, this is compatible with the above if, for large p, $R_p \sim p^\nu$. Thus, if we could compute $\nu(n)$, the correlation length exponent for n-vector models, we would obtain the fractal dimension of the self-avoiding walk as

$$d_\mathrm{f} = \lim_{n\to 0} \frac{1}{\nu(n)}. \tag{6.17}$$

This provides an a priori surprising connection between the non-Markovian problem of self-avoiding walks and the magnetic properties of n-vector models. The renormalization group approach used for n-vector models has provided thus, through the simple continuation to $n = 0$, the geometric indices for a self-avoiding walk, and thus the gyration ratio of polymers.[3]

6.2 Potts model and percolation

Another generalization of the Ising model consists of introducing at each site a 'spin' taking q distinct values, a q-valued 'colour' index. The model is of interest for a number of reasons but, as for the previous example, in the singular limit in which q tends to one, it maps onto the geometric problem of *percolation* on a lattice.[4] We shall restrict ourselves in the following to a two-dimensional lattice. Exact results[5] have shown that there is a critical point for $q \leq 4$ and a first-order transition for $q > 4$. Percolation on a lattice is a random geometric process; one can deal with site percolation or link percolation. Each site or link is occupied with probability p or vacant with probability $(1 - p)$, independently for each site or link. A cluster consists of occupied sites or links, all connected along the lattice links. For small p, there are many small clusters of a few sites or links, but when p increases the size of the clusters increases, and above some threshold p_c there is an infinite cluster with probability one. The divergence of the diameter of the largest cluster when p becomes close to p_c is again a geometric critical point with a characteristic exponent ν. The relationship of this geometric problem to a magnetic phase transition, such as the Potts model, has also been an efficient way of dealing with the problem.

Let us look, for instance, at the link percolation problem. After drawing a configuration, with the given probability law, one obtains a graph $\mathcal{G}$ with a probability $P(\mathcal{G})$. Let us define $N(\mathcal{G}, s, l)$ the number of clusters of $\mathcal{G}$ consisting of s sites and l links. The sum $\sum_s N(\mathcal{G}, s, l)$ is the number of clusters of $\mathcal{G}$ made of l links. Consequently, the number $N(\mathcal{G})$ of links of $\mathcal{G}$ is

$$N(\mathcal{G}) = \sum_l l \sum_s N(\mathcal{G}, s, l). \tag{6.18}$$

[3] For a comparison with experiment see, for instance, M. Daoud, P. G. de Gennes, G. Sarma *et al.*, *Macromolecules*, **8** (1975) 804.

[4] P. W. Kasteleyn and C. M. Fortuin, *J. Phys. Soc. Jpn. Suppl.*, **26** (1969) 11; C. M. Fortuin and P. W. Kasteleyn, *Physica (Utrecht)*, **57** (1972) 536.

[5] R. J. Baxter, *J. Stat. Phys.*, **57** (1989) 1.

The average number per site of clusters with s sites is a number

$$n_s(p) = \frac{1}{N}\sum_{\mathcal{G}} P(\mathcal{G}) \sum_{l} N(\mathcal{G}, s, l), \tag{6.19}$$

in which the probability of the graph $\mathcal{G}$ for a link percolation problem on a lattice with a total number of L links is

$$P(\mathcal{G}) = p^{N(\mathcal{G})}(1-p)^{L-N(\mathcal{G})}. \tag{6.20}$$

The generating function

$$f(p,h) = \sum_{s} n_s(p)\,\mathrm{e}^{-hs} \tag{6.21}$$

allows one to characterize, through derivatives with respect to h at $h = 0$ all the moments of $n_s(p)$.

Let us now return to the q-states Potts model; again we assume that only nearest neighbours interact. It is convenient to normalize the interaction so that the interaction energy of two neighbouring spins σ and σ' takes the value $-J(q-1)$ if the two spins are equal, and $+J$ if they are not equal. The energy of a configuration is thus

$$\beta E = -J\sum_{\langle i,j\rangle}(q\delta_{\sigma_i\sigma_j} - 1). \tag{6.22}$$

The Ising model is nothing but the $q = 2$ model. Exact results by Baxter show that in two dimensions there is a critical point for $q \leq 4$ (and a first-order transition for $q > 4$). In the low-T phase, one may have q distinct pure phases.

In the presence of an external 'field' coupled to one of the colours, for instance the first one, the energy reads

$$\beta E = -J\sum_{\langle i,j\rangle}(q\delta_{\sigma_i\sigma_j} - 1) - h\sum_{i}(q\delta_{\sigma_i,1} - 1). \tag{6.23}$$

We now perform a high-temperature expansion of the partition function. Since the interaction includes a factor β, it is an expansion for small J and small h. The calculation is easier if we note the identities

$$\begin{aligned}\mathrm{e}^{J(q\delta_{\sigma,\sigma'}-1)} &= \mathrm{e}^{J(q-1)}[\mathrm{e}^{-qJ} + (1-\mathrm{e}^{-qJ})\delta_{\sigma,\sigma'}],\\ \mathrm{e}^{h(q\delta_{\sigma,1}-1)} &= \mathrm{e}^{h(q-1)}\,\mathrm{e}^{hq(\delta_{\sigma,1}-1)}.\end{aligned} \tag{6.24}$$

Defining

$$p = 1 - \mathrm{e}^{-qJ}, \tag{6.25}$$

which vanishes with J, one finds

$$Z = \mathrm{e}^{J(q-1)L}\,\mathrm{e}^{h(q-1)N}(1-p)^L \sum_{\mathcal{C}} \prod_{\langle i,j\rangle} \left(1 + \frac{p}{1-p}\delta_{\sigma_i,\sigma_j}\right) \mathrm{e}^{hq\sum_i(\delta_{\sigma_i,1}-1)}, \quad (6.26)$$

in which L is the total number of links of the lattice. Expanding the product, we have the choice, for any link of the lattice, of a factor equal to 1 or to $p/(1-p)$. To represent one term of the expansion of

$$Z' = Z\,\mathrm{e}^{-J(q-1)L}\,\mathrm{e}^{-h(q-1)N}, \quad (6.27)$$

one represents by a bar on the lattice the links for which one has chosen a factor $p/(1-p)$, hence a graph $\mathcal{G}$. One obtains thereby the expression

$$Z' = (1-p)^L \sum_{\mathcal{C}} \sum_{\mathcal{G}} \left(\frac{p}{1-p}\right)^{N(\mathcal{G})} \prod_{\text{links of } \mathcal{G}} \delta_{\sigma,\sigma'}\,\mathrm{e}^{hq\sum_r(\delta_{\sigma_r,1}-1)}. \quad (6.28)$$

Let us now perform the summation over all the configurations of the spins for a given graph $\mathcal{G}$. For a given cluster contained in $\mathcal{G}$ all the sites must be of the same 'colour' (i.e., Potts index). If this is colour number one, the h-dependent factor is simply equal to one; for the $(q-1)$ other colours, a factor e^{-hq} occurs for each spin. The sum over the colours of this cluster with s sites thus yields the factor $[1+(q-1)\,\mathrm{e}^{-hqs}]$. There are $\sum_l N(\mathcal{G},s,l)$ clusters of s sites in the graph, hence a contribution $[1+(q-1)\,\mathrm{e}^{-hqs}]^{\sum_l N(\mathcal{G},s,l)}$ from the clusters of s sites in $\mathcal{G}$. We now collect the contributions of all the clusters with $1, 2, \ldots, s, \ldots$ sites of a given graph and obtain

$$Z' = \sum_{\mathcal{G}} (1-p)^{L-N(\mathcal{G})} p^{N(\mathcal{G})} \prod_{s,l} [1+(q-1)\,\mathrm{e}^{-hqs}]^{\sum_l N(\mathcal{G},s,l)}. \quad (6.29)$$

This expression has been established for integer q, but it may be continued to a continuously varying parameter. In particular, the limit when q tends to one is trivial since

$$\lim_{q\to 1} Z' = \sum_{\mathcal{G}} (1-p)^{L-N(\mathcal{G})} p^{N(\mathcal{G})} = \sum_{\mathcal{G}} P(\mathcal{G}) = 1. \quad (6.30)$$

Expanding now near $q = 1$

$$\prod_{s,l} [1+(q-1)\,\mathrm{e}^{-hqs}]^{\sum_l N(\mathcal{G},s,l)} = 1 + (q-1)\sum_{s,l} \mathrm{e}^{-hs} N(\mathcal{G},s,l) + O((q-1)^2),$$

one obtains

$$\frac{1}{N}\log Z' = (q-1) f(p,h) + O((q-1)^2), \quad (6.31)$$

in which $f(p, h)$ is the generating function of the percolation clusters defined in (6.21). Finally, if one introduces the connectivity of the lattice

$$c = \frac{2L}{N} \tag{6.32}$$

in the thermodynamic limit, in which N tends to infinity with fixed c, one obtains

$$\lim_{N\to\infty} \lim_{q\to 1} \frac{\log Z}{N(q-1)} = \frac{1}{2}cJ + h + f(p, h), \tag{6.33}$$

which shows how to obtain results on the percolation problem by studying the transition of the Potts model for arbitrary q and then taking the singular limit $q \to 1$. For a detailed exposition of the percolation problem see, e.g., D. Stauffer.[6]

[6] D. Stauffer, *Introduction to Percolation Theory*, (London: Taylor and Francis, 1985).

7
Phenomenological description of critical behaviour

Numerous phenomenological descriptions of broken symmetries near the critical point at which they appear have been used in the past. In particular, *Landau theory* provides a simple and general picture, and in many cases it is the starting point for a more microscopic approach. After describing Landau theory I shall compare its results with some exact answers and with experiment. The chapter will end with the notions of *scaling laws and universality.*

7.1 Landau theory

In fact, this is not part of statistical mechanics, which aims to deduce the collective thermodynamic properties from the microscopic laws of interaction. Landau theory is a thermodynamic approach, which starts directly from a construction of the free energy expressed in terms of an *order parameter*. This construction is based on symmetry considerations and assumptions of analyticity.

We begin with a local order parameter $m(\vec{r})$, which will allow us to discriminate between a phase of spontaneously broken symmetry, in which the order parameter will be non-vanishing, and a symmetric phase, in which it vanishes in the absence of an external source. One can think of $m(\vec{r})$ as the local magnetization of a spin system. It may be a simple scalar, or an n-component vector. Indeed the crystalline anisotropies may impose one privileged direction for the magnetization: then one has a uniaxial ferromagnet, i.e., $n = 1$; if there is an 'easy' plane of magnetization one has a planar ferromagnet $n = 2$.

However, the situation can be more complex; for instance in the superconducting transition of liquid ^{3}He, the pairing of two atoms takes place in a relative p-wave ($l = 1, l_z = 1, 0, -1$); the antisymmetry of the wave function (remember that ^{3}He atoms are fermions) imposes a symmetric spin state, i.e., a triplet state $S = 1$, $S_z = 1, 0, -1$. The order parameter carries two indices a and b, one for l_z, the other one for S_z and the order parameter is a 3×3 matrix $m_{a,b}$.

The discussion will be restricted here for simplicity to the n-component vector model, but it should be kept in mind that this is far from exhausting the cases that one finds in nature. The system is supposed to be invariant under a symmetry group $\mathcal{G}$ and m transforms according to a given representation of the group. To make the discussion simpler and more concrete, we shall take the example of $\mathcal{G} = O(n)$, the rotation group in n dimensions, where m is an n-dimensional vector (fundamental representation). (However, in the example of ^{3}He, in which the order parameter is a complex 3×3 matrix, the free energy is invariant under $m \to U_1 m U_2^\dagger$; this group is called $U_3 \times U_3$, i.e., a pair of 3×3 unitary matrices.) Again, we limit ourselves from now on to the group $O(n)$.

The free energy is a functional of this order parameter, called $\Gamma\{m\}$. It is assumed to be invariant under $\mathcal{G}$, i.e., if R is an element of the group, a rotation here, under

$$m(\vec{r}) \to m'_a(\hat{\text{E}}\vec{r}) = \sum_{b=1}^{n} R_{ab} m_b(\vec{r}) \tag{7.1}$$

$$\Gamma\{m'\} = \Gamma\{m\}. \tag{7.2}$$

Note that this is a global symmetry, not a local one: we perform here the same operation R at every point of space. The conjugate variable to m is defined as the functional derivative

$$h_a(\vec{r}) = \frac{\delta \Gamma}{\delta m_a(\vec{r})}. \tag{7.3}$$

In the magnetic example it would be the applied external magnetic field.

Whenever the external applied field h is uniform, i.e., $\vec{r}$-independent, the translational invariance implies that the order parameter is also uniform, $m(\vec{r}) = m$, and

$$\Gamma\{m\} = V\gamma(m), \tag{7.4}$$

in which V is the volume of the sample. In such a case, one verifies that (7.3) reduces to the ordinary derivative

$$h_a = \frac{\partial \gamma}{\partial m_a}. \tag{7.5}$$

In the absence of an external source, i.e., if $h = 0$, the order parameter is an extremum of γ; thermodynamic stability implies that it is a minimum. The assumed rotational invariance implies that $\gamma(m_a)$ is a function of the norm m of the vector m_a. Let us assume that the minimum of γ is located at $m = 0$ for temperatures T greater than some threshold T_c, and that the minimum is no longer at $m = 0$ for

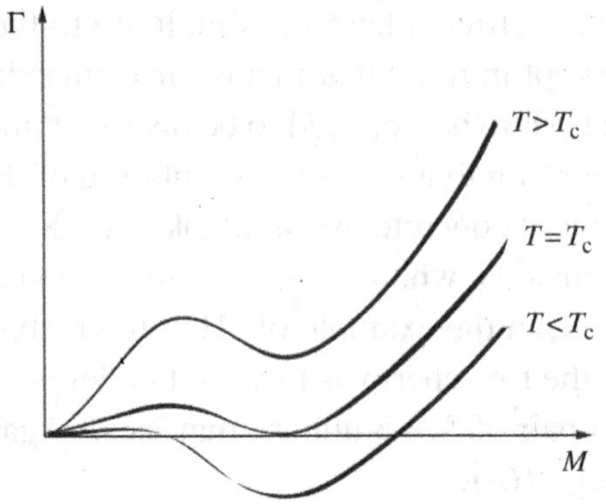

Figure 7.1 First-order transition

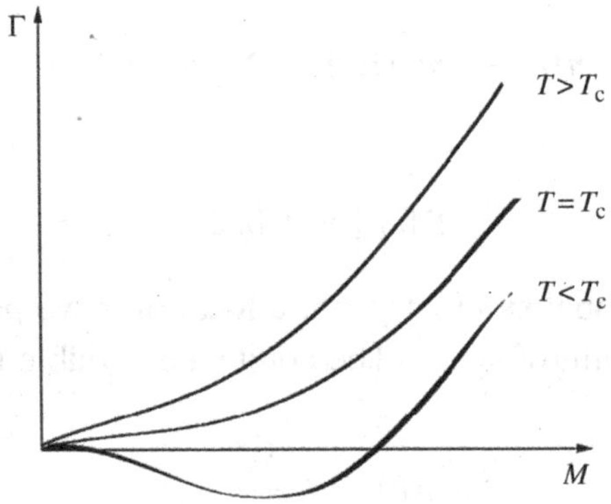

Figure 7.2 Second-order transition

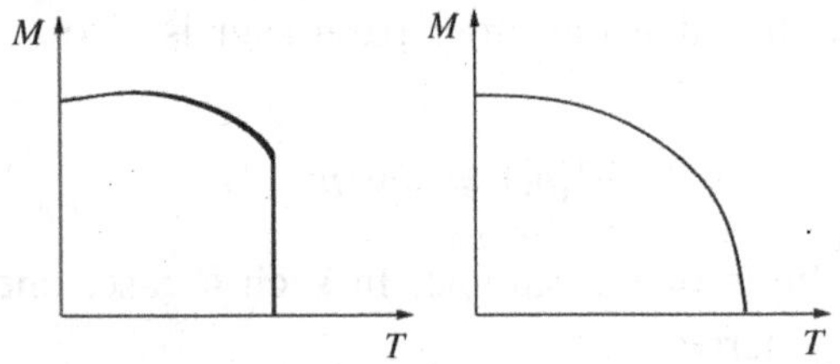

Figure 7.3 The spontaneous magnetization in the two cases

$T < T_c$. Two possibilities are still compatible with these assumptions as one can see in Figs. 7.1 to 7.3.

In the first case, the spontaneous value of the order parameter, i.e., its value for $h = 0$, 'jumps' from zero above T_c to a finite non-vanishing value below T_c: this is called a *first-order transition*. The entropy per unit volume

$$s = -\frac{1}{V}\frac{\partial \Gamma}{\partial T} = -\frac{\partial \gamma}{\partial T} \tag{7.6}$$

in zero external field has a jump at the transition temperature.

In the second case, a spontaneous magnetization also appears below T_c but it vanishes continuously at T_c. This *continuous transition*, often called a second-order transition, presents a *critical point* at T_c and we shall restrict ourselves in the following to this critical case, which presents many intriguing features. The vicinity is marked by a rich singular structure (see Section 2.6) and we shall study what Landau theory predicts for these singularities.

7.2 Landau theory near the critical point: homogeneous case

Since the order parameter m vanishes above T_c and remains small below T_c for T near T_c, it is natural to assume that one can expand the free energy per unit volume in powers of **m**, the norm of the order parameter.

$$\gamma(m, T) = k(T) + \frac{1}{2}a(T)\vec{m}^2 + \frac{1}{4}b(T)(\vec{m}^2)^2 + 0(m^6). \tag{7.7}$$

We assume that $b(T_c)$ is positive and remains positive in the domain of temperatures of interest. (The case of negative b is also of interest but it leads to different physics, such as tricritical points.)

Note that the simplicity of the expansion (7.7) comes from the fact that we have limited ourselves to symmetries under n-dimensional rotations. For an arbitrary group, the expansion should contain all the invariants of increasing orders that one can construct with the order parameter. In particular, for the rotation group there is no cubic invariant. For groups that allow for an invariant of degree three it is easy to see that the transition is of first order. In many cases there may be several invariants of degree four and the discussion which follows must be re-examined.

Exercise

Assume that the order parameter m is an $n \times n$ real symmetric matrix and that the free energy is invariant under $m \to g_1 m g_2^t$, in which g_1 and g_2 belong to the group O_n. Find the quadratic invariant and the quartic ones.

The magnetic field is thus related to the magnetization by (7.5),

$$h_\alpha = a(T)m_\alpha + b\,\vec{m}^2 m_\alpha + O(m^5), \tag{7.8}$$

this equation being the magnetic equation of state (equivalent to the $p = p(V, T)$ for a fluid).

Spontaneous magnetization

This is the solution of (7.8) for $h = 0$. If there are several non-zero solutions, we have to choose the solution that minimizes the free energy.

The obvious solution $m_\alpha = 0$ is a minimum of $\gamma(m)$ for $a(T) > 0$. However, if $a(T) < 0$, this solution is a maximum and the minimum is given by

$$m_{\rm s}^2 = -\frac{a}{b}, \tag{7.9}$$

in which m is the norm of the vector $\vec{m}$.

The critical temperature is thus associated with the vanishing of $a(T)$ and its change of sign. The simplest assumption is to take

$$a(T) = a_0 t + O(t^2), \tag{7.10}$$

in which

$$t = \frac{(T - T_{\rm c})}{T_{\rm c}}. \tag{7.11}$$

Then the spontaneous magnetization $m_{\rm s}(t)$ is given by

$$m_{\rm s}(T) = \begin{cases} 0 & \text{if } t > 0 \\ \left(-\frac{a_0}{b} t\right)^{1/2} & \text{if } t < 0. \end{cases} \tag{7.12}$$

This behaviour may be characterized by a *critical exponent* β,

$$m_{\rm s} \simeq (-t)^\beta, \tag{7.13}$$

with

$$\beta = \frac{1}{2}. \tag{7.14}$$

Note that if we had kept the terms of degree greater than four in the expansion (7.7) this would not have affected the value of β. And similarly, if we had kept the term of order t^2 in the expansion of $a(T)$.

If the norm of the spontaneous magnetization is fixed by these equations, its direction remains arbitrary, on a sphere if $n > 1$ or ± 1 if $n = 1$. Again it is through boundary conditions or by the direction imposed by a vanishingly small external field that a pure state, of given magnitude and orientation for $\vec{m}_{\rm s}$ may be found. We shall call $\vec{e}$ the unit vector along $\vec{m}_{\rm s}$:

$$\vec{m}_{\rm s} = m_{\rm s}\vec{e}. \tag{7.15}$$

Susceptibility

The vicinity of the critical point is also marked by a strong response of the magnetization to an applied external field. Let us examine the susceptibility when the applied field tends to zero:

$$\chi_{\alpha,\beta} = \left.\frac{\partial m_\alpha}{\partial h_\beta}\right|_{h=0}. \tag{7.16}$$

The inverse of χ (in the matrix sense) is

$$\begin{aligned}\frac{\partial h_\alpha}{\partial m_\beta} &= (a + bm^2)\delta_{\alpha,\beta} + 2bm_\alpha m_\beta \\ &= (a + bm^2)(\delta_{\alpha,\beta} - e_\alpha e_\beta) + (a + 3bm^2)e_\alpha e_\beta.\end{aligned} \tag{7.17}$$

On the last expression one sees that this matrix has one 'longitudinal' eigenvalue $(a + 3bm^2)$ and $(n - 1)$ equal 'transverse' eigenvalues $a + bm^2$. Therefore,

$$\chi_{\alpha,\beta} = \frac{1}{(a + bm^2)}(\delta_{\alpha,\beta} - e_\alpha e_\beta) + \frac{1}{(a + 3bm^2)}e_\alpha e_\beta, \tag{7.18}$$

and we can define a longitudinal susceptibility,

$$\chi_{\mathrm{L}} = \frac{1}{(a + 3bm^2)}, \tag{7.19}$$

and a transverse one

$$\chi_{\mathrm{T}} = \frac{1}{(a + bm^2)}. \tag{7.20}$$

Let us look at what happens if we let the external field vanish.

Longitudinal susceptibility Above T_c, since there is no spontaneous magnetization,

$$\chi_{\mathrm{L}} = \frac{1}{a(T)} = \frac{\chi_0}{t}, \tag{7.21}$$

i.e., a linear divergence,

$$\chi_{\mathrm{L}} \simeq \frac{1}{t^\gamma} \quad \text{with} \quad \gamma = 1. \tag{7.22}$$

Below T_c, we have $m_s^2 = -a/b$ and

$$\chi_{\mathrm{L}} = \frac{1}{-2a(T)} = -\frac{\chi_0}{2t}, \tag{7.23}$$

again a linear divergence

$$\chi_{\mathrm{L}} \simeq \frac{1}{(-t)^{\gamma'}} \quad \text{with} \quad \gamma' = \gamma = 1. \tag{7.24}$$

Transverse susceptibility For $n > 1$ we have $(n - 1)$ transverse directions. Above T_c in zero field

$$\chi_T = \chi_L \simeq \frac{1}{t} \tag{7.25}$$

but below T_c, since $(a + bm^2)$ vanishes with the field, the transverse susceptibility is infinite! This phenomenon, related to the presence of Goldstone modes, will be discussed in Chapter 13.

Specific heat

In zero field, the free energy per unit volume (7.7) is equal to

$$\gamma(T) = k(T) \tag{7.26}$$

above T_c, but below T_c

$$\gamma(T) = k(T) - \frac{a^2}{4b}. \tag{7.27}$$

The free energy is thus continuous at T_c (since $a(T)$ vanishes), its first derivative (equal to minus the entropy per unit volume) also vanishes (since the derivative of a^2 vanishes). But the specific heat per unit volume, equal to $-T\frac{\partial^2\gamma}{\partial T^2}$ has a discontinuity. It is customary to introduce two critical exponents α and α' to characterize the singularity of the specific heat

$$C = C_0^+ t^\alpha \quad \text{for} \quad t > 0; \qquad C = C_0^-(-t)^{\alpha'} \quad \text{for} \quad t < 0 \tag{7.28}$$

but in the Landau theory

$$\alpha = \alpha' = 0 \tag{7.29}$$

and the singularity is a step discontinuity.

Critical isotherm

Another critical exponent characterizes the critical isotherm, namely the relation between the external field and the magnetization at the critical temperature T_c. Then the relation (7.7) between the magnitude of the field and that of the magnetization takes the form

$$h = bm^3 + O(m^5), \tag{7.30}$$

i.e.,

$$h \simeq m^\delta \quad \text{with} \quad \delta = 3. \tag{7.31}$$

7.3 Landau theory and spatial correlations

We want now to study the range of correlations in this system, for instance, we would like to know how a change of the magnetic field at site $\vec{r}'$ affects the magnetization at site $\vec{r}$. This means that we have to extend the Landau theory to allow for spatially varying order parameters. Up to now we have worked with a lattice, but we shall now consider continuous space. This may be achieved, if we started with a lattice model, by a coarse graining of the magnetization: we consider a small cell that is large compared with the lattice spacing a, but small compared with the correlation length (I shall define it below and verify that it is very large near the critical point). From now on, everything that I discuss will be limited to the cases in which the correlation length is much larger than the lattice spacing. So the coarse-grained magnetization $\vec{m}(\vec{r})$ is the average of the lattice magnetization in some neighbourhood of $\vec{r}$.

Remark

This coarse graining has the effect of supressing the quantum mechanical properties of the spins. If

$$\vec{m}(\vec{r}) = \frac{1}{p}\vec{S}_i, \tag{7.32}$$

in which p is the number of spins in the coarse graining cell surrounding $\vec{r}$ then, for any spin we have

$$[S_i^\alpha, S_j^\beta] = \mathrm{i}\hbar\delta_{i,j}\epsilon^{\alpha\beta\gamma}S_j^\gamma \tag{7.33}$$

but for $\vec{m}(\vec{r}) = \frac{1}{p}\sum_i \vec{S}_i$ one finds

$$[m^\alpha, m^\beta] = \frac{1}{p}\epsilon^{\alpha\beta\gamma}m^\gamma. \tag{7.34}$$

This factor $1/p$ shows the reduction of quantum effects and explains why we can limit ourselves here to classical commuting order parameters. It is again the fact that the correlation length is large that allows coarse graining with a relatively large p and explains that critical points are mostly classical (critical points at zero temperature are an exception to this statement).

The free energy is thus a function of $m^\alpha(\vec{r})$, in which $\vec{r}$ is a d-dimensional vector and the order parameter has n components ($\alpha = 1, \ldots, n$). The expansion (7.7) was based on the fact that near T_c the order parameter was small. Furthermore, since we expect long-range correlations, the order parameter should also vary slowly in space and we can assume that we can also expand in powers of the gradients of the order parameter:

$$\Gamma\{m(\vec{r})\} = \Gamma_0(T) + \int \mathrm{d}^d\vec{r}\left[\frac{1}{2}a(T)\mathbf{m}^2 + \frac{1}{4}b(\mathbf{m}^2)^2 + O(m^6) + \frac{1}{2}\sum_{a=1}^{d}\sum_{\alpha=1}^{n}\left(\frac{\partial m_\alpha}{\partial r_a}\right)^2 + O(\nabla m^4)\right]. \tag{7.35}$$

Note that rotational invariance in the n-dimensional internal space of the order parameter implies that $\left(\frac{\partial m_\alpha}{\partial r_a}\right)^2$ is independent of α, but it may vary with a. We have implicitly assumed that by an appropriate rescaling of space $r_a \to \lambda_a r_a$, we have made all these coefficients equal to $\frac{1}{2}$ of (7.35).

Of course, if the order parameter is constant in space, (7.35) reduces to the previous simple form (7.7), but now we are in a position to study new effects. Assume that we wish to determine the influence on $m_\alpha(\vec{r})$ of a change of the external field at $\vec{r}'$:

$$G_{\alpha,\beta}(\vec{r},\vec{r}') = \frac{\delta m_\alpha(\vec{r})}{\delta h_\beta(\vec{r}')}. \tag{7.36}$$

It is easier to calculate first its inverse (in the matrix and kernel sense) and we first compute

$$h_\alpha(\vec{r}) = \frac{\delta\Gamma}{\delta m_\alpha(\vec{r})}, \tag{7.37}$$

since from (7.35) we find

$$h_\alpha(\vec{r}) = (a + b\mathbf{m}^2 - \nabla^2)m_\alpha(\vec{r}). \tag{7.38}$$

It is thus easier to compute the inverse of G, namely

$$\frac{\delta h_\alpha(\vec{r})}{\delta m_\beta(\vec{r}')}. \tag{7.39}$$

We now assume that, after taking this variation, the external field is reduced to zero. The magnetization is thus a constant vector

$$m_\alpha(\vec{r}) = m_\mathrm{s} e_\alpha, \tag{7.40}$$

in which m_s is given by (7.9). Then

$$F_{\alpha\beta}(\vec{r}-\vec{r}') = \frac{\delta h_\alpha(\vec{r})}{\delta m_\beta(\vec{r}')} = \delta_{\alpha\beta}(a + bm_\mathrm{s}^2 - \nabla^2)\delta(\vec{r}-\vec{r}') + e_\alpha e_\beta 2bm_\mathrm{s}^2\delta(\vec{r}-\vec{r}'). \tag{7.41}$$

The translational invariance of this problem leads to the solution: we define

$$g_{\alpha\beta}(q) = \int \mathrm{d}^d r \mathrm{e}^{\mathrm{i}\vec{q}\cdot\vec{r}} G_{\alpha,\beta}(\vec{r}) \tag{7.42}$$

and

$$f_{\alpha\beta}(q) = \int \mathrm{d}^d r \mathrm{e}^{\mathrm{i}\vec{q}\cdot\vec{r}} F_{\alpha,\beta}(\vec{r}) \tag{7.43}$$

and the equation (7.41) gives immediately

$$\begin{aligned} f_{\alpha\beta}(q) &= \delta_{\alpha\beta}(a + bm_{\mathrm{s}}^2 + q^2) + e_\alpha e_\beta 2bm_{\mathrm{s}}^2 \\ &= (\delta_{\alpha\beta} - e_\alpha e_\beta)(a + bm_{\mathrm{s}}^2 + q^2) + e_\alpha e_\beta(a + 3bm_{\mathrm{s}}^2 + q^2). \end{aligned} \tag{7.44}$$

The $n \times n$ matrix $g_{\alpha\beta}(q)$ is the inverse in the matrix sense of $f_{\alpha\beta}(q)$. The term $e_\alpha e_\beta$ is the matrix element of the longitudinal propagator P_{L} on the direction of $\mathbf{e}$, whereas $(\delta_{\alpha\beta} - e_\alpha e_\beta)$ is the transverse projector P_{T}, which projects onto the $(n-1)$ directions orthogonal to $\mathbf{e}$. Therefore, we have immediately the inverse

$$g_{\alpha\beta}(q) = (\delta_{\alpha\beta} - e_\alpha e_\beta)\frac{1}{(a + bm_{\mathrm{s}}^2 + q^2)} + e_\alpha e_\beta \frac{1}{(a + 3bm_{\mathrm{s}}^2 + q^2)}. \tag{7.45}$$

- Above T_{c} in zero field, $m_{\mathrm{s}} = 0$ and thus the longitudinal and the transverse parts are equal

$$g_{\mathrm{L}}(q) = g_{\mathrm{T}}(q) = \frac{1}{a + q^2}. \tag{7.46}$$

Consequently,

$$G_{\mathrm{L}}(\vec{r}) = G_{\mathrm{T}}(\vec{r}) = \int \frac{\mathrm{d}^d q}{(2\pi)^d} \frac{\mathrm{e}^{-\mathrm{i}\vec{q}\cdot\vec{r}}}{q^2 + a(T)}. \tag{7.47}$$

The vector $\vec{r}$ stands here for the difference $\vec{r} - \vec{r}'$. The asymptotic behaviour of this integral is well-known (see the appendix at the end of this chapter). Defining the correlation length

$$\xi = \frac{1}{\sqrt{a(T)}} \tag{7.48}$$

gives an exponential fall-off of the correlations of the form

$$G(r) \sim \frac{\mathrm{e}^{-\frac{r}{\xi}}}{r^{(d-1)/2}}. \tag{7.49}$$

Since $a(T)$ tends to zero at T_{c}, there is indeed a divergent correlation length with the behaviour

$$\xi \simeq \frac{\xi_0^{(+)}}{t^\nu} \quad \text{with} \quad \nu = \frac{1}{2}. \tag{7.50}$$

- At T_c

$$G_L(\vec{r}) = G_T(\vec{r}) = \int \frac{d^d q}{(2\pi)^d} \frac{e^{-i\vec{q}\cdot\vec{r}}}{q^2} \tag{7.51}$$

and on purely dimensional grounds one sees that $G(r)$ falls off as $1/r^{(d-2)}$, namely

$$G(r) \sim \frac{1}{r^{d-2+\eta}} \quad \text{with} \quad \eta = 0. \tag{7.52}$$

- Below T_c the longitudinal and transverse responses are different; replacing m_s^2 by its value, one finds

$$G_L(\vec{r}) = \int \frac{d^d q}{(2\pi)^d} \frac{e^{-i\vec{q}\cdot\vec{r}}}{q^2 - 2a(T)}. \tag{7.53}$$

As before, we now have an exponential fall-off with a correlation length

$$\xi = \frac{1}{\sqrt{-2a(T)}}, \tag{7.54}$$

i.e.,

$$\xi \simeq \frac{\xi_0^{(-)}}{(-t)^{\nu'}} \quad \text{with} \quad \nu' = \frac{1}{2}. \tag{7.55}$$

However, the transverse response is

$$G_T(\vec{r}) = \int \frac{d^d q}{(2\pi)^d} \frac{e^{-i\vec{q}\cdot\vec{r}}}{q^2}, \tag{7.56}$$

which falls off only as the power $1/r^{(d-2)}$.

7.4 Transitions without symmetry breaking: the liquid–gas transition

Everything up to now relied on a free energy, invariant under some symmetry. However, if we look at the transition from a fluid ($T > T_c$) to a phase where liquid and vapour can coexist ($T < T_c$), one does not see any symmetry breaking. The order parameter would be proportional to the difference of volume per particle between the two phases, and one can consider an expansion in powers of this difference for the pressure:

$$P = P_0 + am + bm^2 + cm^3. \tag{7.57}$$

A simple translation on m may be used to cancel the m^2 term and one finds

$$P - P_c = \tilde{a}m + \tilde{b}m^3, \tag{7.58}$$

exactly as for the relation h versus m in an $n = 1$ magnetic system.

Therefore, accidentally in a way, the critical properties of the liquid–gas to fluid transition are identical to $n = 1$ magnets. However, the absence of symmetry manifests itself in the corrections to the leading critical behaviour. Indeed, for the magnetic equation of state, the Z_2 symmetry prevents the occurrence of a term proportional to m^4 in the magnetic equation of state, whereas such a term is allowed in (7.58).

7.5 Thermodynamic meaning of $\Gamma\{m\}$

Let us return for a while to statistical mechanics. There we know that the free energy $F\{h\}$ is obtained (or rather obtainable in some simple cases) from the logarithm of the partition function. If, for simplicity, we take a homogeneous field h then the free energy per unit volume is

$$f(h) = -\frac{1}{V}\frac{1}{kT}\log Z \tag{7.59}$$

and the magnetization per unit volume is

$$m_\alpha = -\frac{\partial f}{\partial h_\alpha}. \tag{7.60}$$

Imagine that we solve this last relation for $h = h(m)$, then the free energy per unit volume $\gamma(m)$ is defined as

$$\gamma(m) = f(h) + \mathbf{h}\cdot\mathbf{m}. \tag{7.61}$$

The transformation $f \to \gamma$ (Legendre transformation) is symmetric, as one can verify by taking the derivative of both sides with respect to m:

$$\frac{\partial\gamma}{\partial m_\alpha} = \sum_\beta\left(\frac{\partial f}{\partial h_\beta}\frac{\partial h_\beta}{\partial m_\alpha} + m_\beta\frac{\partial h_\beta}{\partial m_\alpha}\right) + h_\alpha. \tag{7.62}$$

The cancellation of the first two terms leads to

$$h_\alpha = \frac{\partial\gamma}{\partial m_\alpha}. \tag{7.63}$$

Let us look more closely at what this implies in the simple case $n = 1$. Let us imagine that we have computed $f(h)$ in an Ising-like model. The resulting $f(h)$ is even in h, by symmetry. However, its derivative as $h \to +0$ is non-zero below T_c, since it is $-m_s$; its derivative when $h \to -0$ is opposite, it is $+m_s$. Therefore, $f(h)$ exhibits a cusp at $h = 0$. The function $f(h)$ is convex, i.e., $f''(h) \leq 0$ (GKS inequality). This implies that, for $h > 0$, m increases from m_s. If we plot the resulting $h(m)$, and by integration $\gamma(m)$ one sees that the 'true' $h(m)$ and $\gamma(m)$ are not given below T_c by a formula such as $h = am + bm^3$ and $\gamma(m) = \frac{1}{2}am^2 + \frac{1}{4}bm^4$ but for γ by the convex envelope of Landau formula, and for h by Maxwell's construction: i.e., a coexistence straight line rather than a loop, erasing two pieces of equal area (Figs 7.4 to 7.7).

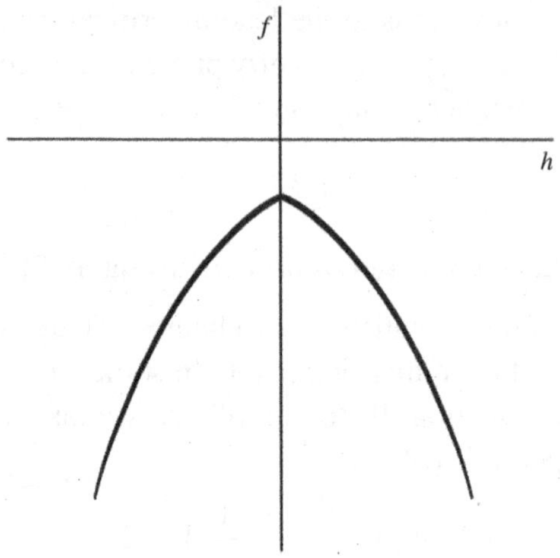

Figure 7.4 The free energy as a function of the applied field

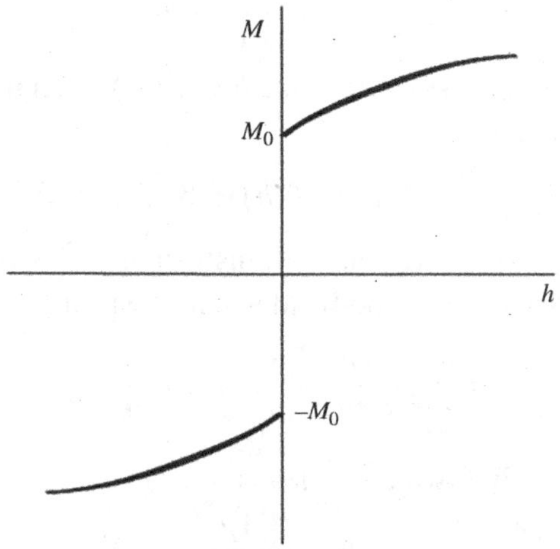

Figure 7.5 The magnetization versus the applied field

7.6 Universality

The critical behaviour given by Landau theory is completely independent of the specific system under consideration; the results that we have found are independent of the dimension of space, of the symmetry group, etc. However, Table 7.1 shows that they do not agree with the exact results of the two-dimensional Ising models; one could argue that two dimensions are special and that for real three-dimensional experimental systems this simple theory applies. But the experimental results show that, if Landau theory is qualitatively simple and powerful, in practice it is not quite right.

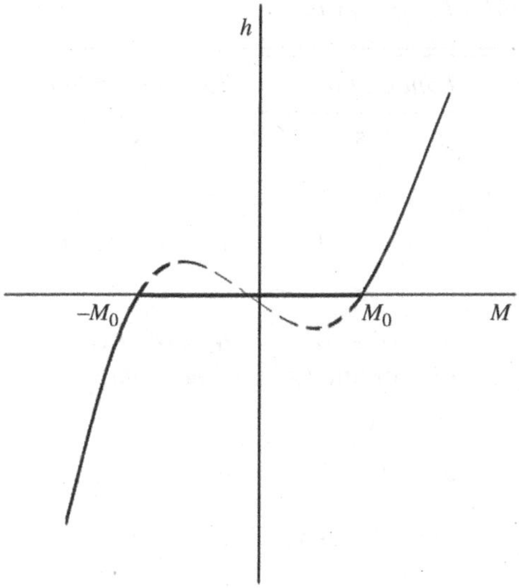

Figure 7.6 The field versus the magnetization

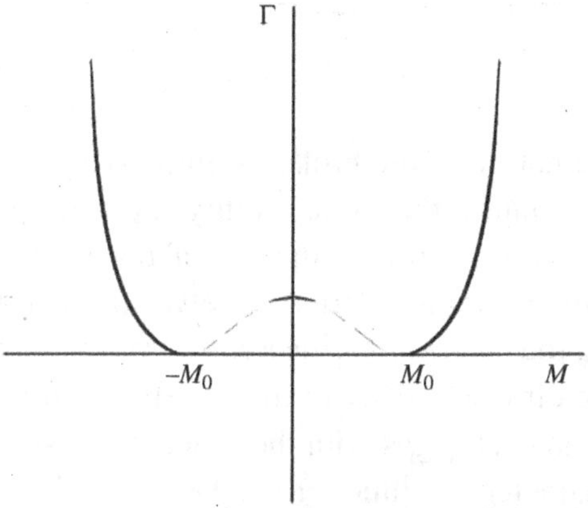

Figure 7.7 The convexity of the free energy as a function of the magnetization

Given this failure, one could have thought that the true behaviour is specific to each system: every species of molecules having a liquid–gas critical point would have its own behaviour; every magnetic system would be different. Indeed, there is little in common between helium, whose liquid–gas critical point is $T_c \simeq 5.2\,\mathrm{K}$ and CO_2 where $T_c \simeq 304\,\mathrm{K}$. Nevertheless, the accumulation of precise experiments has revealed that all the liquid–gas transitions have the same set of exponents, the same equation of state if we take reduced variables $(P/P_c, \rho/\rho_c, T/T_c)$, the same correlation functions as function of r/ξ. Furthermore, this same set of critical properties is valid for all the $n = 1$ systems; those in which the order parameter is a simple scalar, uniaxial ferromagnets, the critical point of demixion of two fluids, etc.

Table 7.1 *Table of critical exponents*

Definition	*Landau theory*	*Two-dimensional Ising*	$d = 3$ $(n = 1)$
Spontaneous magnetization $m_s \sim (-t)^\beta$	$\beta = 0,5$	$\beta = 1/8$	$\beta \simeq 0.325$
Susceptibility ($h = 0$) $\chi \sim t^\gamma$ for $t > 0$, $\chi \sim (-t)^{\gamma'}$ for $t < 0$	$\gamma = \gamma' = 1$	$\gamma = \gamma' = 7/4$	$\gamma = \gamma' \simeq 1.240$
Specific heat $C \sim t^{-\alpha}$ for $t > 0$ $C \sim (-t)^{-\alpha'}$ for $t < 0$	$\alpha = \alpha' = 0$ (discontinuity)	$\alpha = \alpha' = 0$ (log–divergence)	$\alpha \simeq \alpha' \simeq 0.11$
Critical isotherm $h(T_c) \sim m^\delta$	$\delta = 3$	$\delta = 15$	$\delta \simeq 5.2$
Correlations ($T \neq T_c$) $\xi \sim t^\nu$ for $t > 0$ $\xi \sim (-t)^{\nu'}$ for $t < 0$	$\nu = \nu' = 0.5$	$\nu = \nu' = 1$	$\nu \simeq \nu' \simeq 0.63$
Correlations ($T = T_c$) $G(r) \sim 1/r^{d-2+\eta}$	$\eta = 0$	$\eta = 1/4$	$\eta \simeq 0.03$

However, when the nature of the broken symmetry changes, the critical properties also change. For instance, the susceptibility exponent $\gamma \simeq 1.24$ for $n = 1$, but it is close to 1.43 for $n = 3$. Therefore, the critical behaviour is *universal* but within a given class of symmetry. If we restrict ourselves to the rotation group acting on n-component vectors, the *classes of universality* are simply labelled by n. They are also dependent on the dimensionality of the system: contrary to the Landau theory the critical behaviour also changes with the space dimensionality.

This universal character, common to a large number of different systems, remained a puzzle until the renormalization group approach of K. Wilson allowed one to understand the origin of this extraordinary phenomenon.

7.7 Scaling laws

There are many kinds of scaling law in this problem. First of all, a singularity of some physical quantity x of the type $x \simeq t^{-\kappa}$ gives rise to an elementary scaling law: the result of a measurement in the range $10^{-p} < t < 10^{-p+1}$ can be superposed to the range $10^{-p-1} < t < 10^{-p}$ by a simple rescaling of x. In other words, the system is self-similar at different scales, the critical exponent being the characteristic of the rescaling.

Furthermore, it has been found that the nine exponents defined in Table 7.1 satisfy seven relations: only two exponents are independent. These relations have often been derived as inequalities, but the renormalization group approach, as well as the exact solutions to models, do conclude that they are equalities. They may be written

$$\alpha = \alpha', \qquad \gamma = \gamma', \qquad \nu = \nu'$$
$$\gamma = \nu(2-\eta), \qquad \alpha + 2\beta + \gamma = 2, \qquad \delta = 1 + \frac{\gamma}{\beta}$$
$$\alpha = 2 - \nu d. \tag{7.64}$$

Only the last equality involves the dimension of space and we shall see that it is valid only for $d \leq 4$. The exponents of the two-dimensional Ising model satisfy this relation, and the Landau exponents too... provided $d = 4$. The renormalization group approach will explain why two exponents only, namely η, related to the dimension of the field, and ν, of the energy, suffice.

These relations (7.64) are also called scaling laws because they reflect some deeper true scaling property. For instance, one finds that the magnetic equation of state, i.e., the relation which links (h, m, t) is invariant under a rescaling $t \to \lambda t$, $m \to \lambda^{\beta} m$, $h \to h\lambda^{\beta\delta}$. In other words, one has only one independent variable, not two, and one may write, for instance,

$$\frac{h}{m^{\delta}} = \varphi\left(\frac{t}{m^{1/\beta}}\right). \tag{7.65}$$

This reduction to one variable has been well verified experimentally and the function φ itself is universal. However, we know that above T_c, the magnetization m vanishes linearly with the applied field h. It is a simple exercise to verify that this implies the function $\varphi(x) \sim x^{-\beta(\delta-1)}$ for x large, and that the susceptibility diverges when $t \to 0$ as $t^{-\beta(\delta-1)}$, hence the relation $\gamma = \beta(\delta - 1)$.

Let us now examine the scaling properties of the correlation function. There again the (longitudinal) response function $G(\vec{r}, t)$ is found to be of the simple form

$$G(\vec{r}, t) = \xi^{p} g\left(\frac{r}{\xi}\right), \tag{7.66}$$

with a universal function $g(x)$. The limit $T \to T_c$, at which $G(\vec{r}, 0) \sim \frac{1}{r^{d-2+\eta}}$, implies that $g(x) \sim x^{p}$ when $x \to 0$ and that $p = -(d - 2 + \eta)$. The magnetic susceptibility is the integral over space of $G(\vec{r}, t)$. It is thus proportional to ξ^{p+d}, i.e., to $t^{-\nu(p+d)}$, hence showing again that the relation $\gamma = \nu(2 - \eta)$ is simply a consequence of the scaling law for the correlation function.

Appendix Large-distance behaviour of the free propagator

We return to

$$G(\vec{r}) = \int \frac{\mathrm{d}^d q}{(2\pi)^d} \frac{\mathrm{e}^{-\mathrm{i}\vec{q}\cdot\vec{r}}}{q^2 + \mu^2}$$

to find its behaviour at large distance. If we write

$$\frac{1}{q^2 + \mu^2} = \int_0^\infty \mathrm{d}t \, \mathrm{e}^{-t(q^2+\mu^2)},$$

the integrations over $\vec{q}$ are Gaussian and we find

$$G(r) = \frac{1}{(4\pi)^{d/2}} \int_0^\infty \frac{\mathrm{d}t}{t^{d/2}} \mathrm{e}^{-\left(\frac{r^2}{4t} + t\mu^2\right)}.$$

If we rescale $t \to \frac{r}{2\mu} t$, we obtain

$$G(r) = \frac{1}{(4\pi)^{d/2}} \left(\frac{r}{2\mu}\right)^{1-d/2} \int_0^\infty \frac{\mathrm{d}t}{t^{d/2}} \mathrm{e}^{-\frac{r\mu}{2}\left(t + \frac{1}{t}\right)}.$$

For μr large the integration over t is governed by the saddle-point t_0 solution of $\frac{\mathrm{d}}{\mathrm{d}t}(t + 1/t) = 0$, i.e., by the point $t_0 = 1$ on the integration contour. If we write $t = 1 + x$, then $t + 1/t = 2 + x^2 + O(x^3)$ and the Gaussian integration over x gives the result

$$G(r) \sim \frac{\mathrm{e}^{-\mu r}}{r^{(d-1)/2}}.$$

8
Mean field theory

Landau theory was purely thermodynamic, not microscopic. I want to prove in this chapter that in statistical physics one may recover Landau theory through a standard and simple approximation: mean field theory, in which the interactions between the constituents are replaced by an average which acts independently on every degree of freedom.

8.1 Weiss 'molecular field'

Pierre Curie's experiments on ferromagnetism (1895) have demonstrated the existence of a paramagnetic–ferromagnetic transition, and shown that at high temperature the magnetic susceptibility falls off as $1/T$. Let us show that a model of independent spins is sufficient to recover Curie's $1/T$ law.

If we take simply an energy of interaction of the spins with an external magnetic field;

$$E(C) = -h \sum \sigma_i \tag{8.1}$$

and the spins are independent. The probability that one spin takes the value $\sigma = \pm 1$ is simply

$$p(\sigma) = \frac{1}{2\cosh \beta h}\, \mathrm{e}^{\beta h \sigma} \tag{8.2}$$

and the magnetization $m = \langle \sigma \rangle$ is given by

$$m = \sum_{\sigma} \sigma p(\sigma) = \tanh \beta h. \tag{8.3}$$

For large T this reduces to $m \sim \beta h$ and the susceptibility $\partial m / \partial h$ is indeed proportional to $1/T$.

A few years later, Weiss modified this simple picture by postulating an additional external field acting on a given spin (at the time spin was unknown; it was thought of as a magnetic dipole). This 'molecular field' results from the field created by the other moments acting on the one under consideration. Therefore, this additional field is larger if the magnetization in the system is larger; so Weiss postulated the existence of an effective field acting on each spin of the form

$$h_{\text{eff}} = h + \alpha m. \tag{8.4}$$

If we use this effective field in (8.3) we obtain a self-consistent equation

$$m = \tanh(\beta h_{\text{eff}}) = \tanh \beta(h + \alpha m). \tag{8.5}$$

One verifies easily that in zero field this equation has only the paramagnetic solution $m = 0$ if $\alpha\beta < 1$ (Fig. 8.1), but for $\alpha\beta > 1$ there are three solutions (Fig. 8.2) and until we compute a free energy we cannot really conclude that the solutions with $m \neq 0$ are favoured. This is why we have to go beyond this simple picture.

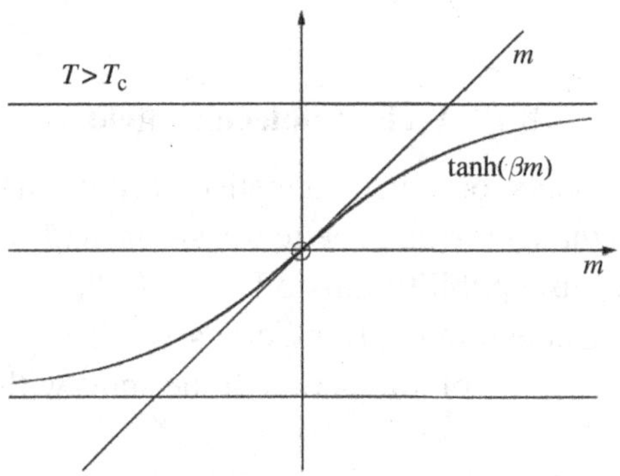

Figure 8.1 Weiss equation $T > T_c$

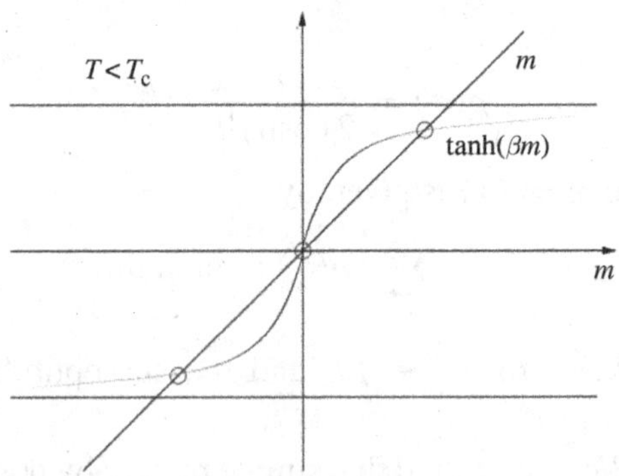

Figure 8.2 Weiss equation $T < T_c$

In the paramagnetic phase $\alpha\beta < 1$, m vanishes with h; therefore, for small h one has simply

$$m = \beta(h + \alpha m) \qquad \text{i.e.,} \quad \frac{m}{h} = \frac{1}{kT - \alpha}. \tag{8.6}$$

The susceptibility satisfies the $1/T$ law at high temperature, but it diverges at a critical temperature $kT_c = \alpha$.

This self-consistent approach, i.e., m determined as a result of a field depending on m itself, provides an interesting picture at this stage but is not a microscopic theory.

8.2 Mean field theory: the variational method

How can one recover this simple molecular field picture within statistical physics? Let us take a spin model, such as

$$E(\mathcal{C}) = -\frac{1}{2}\sum_{i,j} J_{ij}\sigma_i\sigma_j - \sum_i h_i\sigma_i, \tag{8.7}$$

in which we can consider a general interaction J_{ij} (invariant though by lattice translations). An easy way of deriving the mean field equations is to write

$$\sigma_i\sigma_j = (\sigma_i - \langle\sigma_j\rangle)(\sigma_j - \langle\sigma_j\rangle) + \sigma_i\langle\sigma_j\rangle + \sigma_j\langle\sigma_i\rangle - \langle\sigma_i\rangle\langle\sigma_j\rangle \tag{8.8}$$

and to drop the first of the four terms in the r.h.s., which is quadratic in the fluctuations of the spins with respect to their average. This leads immediately to a simple independent-spins problem, but we shall choose a slightly different route, which has the advantage of allowing us to compute the correlation of distant sites. Led by the Curie–Weiss picture, we want to replace this $E(\mathcal{C})$ by an energy of independent spins in some unknown field

$$E_0(\mathcal{C}) = -\sum_i x_i\sigma_i, \tag{8.9}$$

in which the yet unspecified parameters x_i will be fixed by a variational principle, which will make the effect of E_0 as close as possible to that of E.

Notation

$$\langle A\rangle = \frac{1}{Z}\sum_{\mathcal{C}} A\,\mathrm{e}^{-\beta E(\mathcal{C})}$$

$$\langle A\rangle_0 = \frac{1}{Z_0}\sum_{\mathcal{C}} A\,\mathrm{e}^{-\beta E_0(\mathcal{C})}, \tag{8.10}$$

then

$$\frac{Z}{Z_0} = \frac{\sum_{\mathcal{C}} e^{-\beta E(\mathcal{C})}}{\sum_{\mathcal{C}} e^{-\beta E_0(\mathcal{C})}} = \langle e^{-\beta(E-E_0)} \rangle_0. \tag{8.11}$$

We now use the convexity of the exponential to establish the inequality (the proof is simple; it is left to the reader)

$$\langle e^X \rangle \geq e^{\langle X \rangle}. \tag{8.12}$$

This leads to

$$\log \frac{Z}{Z_0} \geq -\beta \big(\langle E(\mathcal{C}) - E_0(\mathcal{C}) \rangle_0 \big) \tag{8.13}$$

and for the free energy

$$F \leq F_0 + \langle E - E_0 \rangle_0. \tag{8.14}$$

The inequality (8.14) is valid for any choice of the parameters x_i. Therefore, we will choose these parameters in a way that minimizes the r.h.s. of (8.14) to be as close as possible to F within the class of independent-spin Hamiltonians.

The mean field free energy is thus defined as

$$F^{\text{m.f.}} = \inf_{x_i} [F_0 + \langle E - E_0 \rangle_0]. \tag{8.15}$$

We have thus to calculate the terms of the r.h.s. of (8.15) and minimize afterwards. The calculations are straightforward:

$$Z_0 = \prod_i 2 \cosh \beta x_i \tag{8.16}$$

$$\langle \sigma_i \rangle_0 = \frac{1}{\beta} \frac{\partial \log Z_0}{\partial x_i} = \tanh(\beta x_i) \tag{8.17}$$

and since, with the Boltzmann weight given by E_0, the spins are independent

$$\langle \sigma_i \sigma_j \rangle_0 = \langle \sigma_i \rangle_0 \langle \sigma_j \rangle_0. \tag{8.18}$$

Then one obtains

$$\begin{aligned} F^{\text{m.f.}} &= \inf_x [F_0 + \langle E - E_0 \rangle_0] \\ &= \inf_x \Bigg[-\frac{1}{\beta} \sum_i \log (2 \cosh \beta x_i) \\ &\qquad - \sum_i (h_i - x_i) \tanh \beta x_i - \frac{1}{2} \sum_{i,j} J_{ij} \tanh \beta x_i \tanh \beta x_j \Bigg]. \end{aligned} \tag{8.19}$$

The minimization with respect to x_i gives

$$x_i - h_i = \sum_j J_{ij} \tanh \beta x_i, \tag{8.20}$$

an equation which shows that $(x - h)$ is nothing but the additional molecular field of Weiss. In this mean field theory one can calculate the magnetization

$$m_i = -\frac{\partial F^{\text{m.f.}}}{\partial h_i}. \tag{8.21}$$

Some care is needed since $F^{\text{m.f.}}$ depends on h explicitly, and through the x_i, but the stationarity conditions give the simple result

$$m_i = \tanh \beta x_i. \tag{8.22}$$

To compare with Landau theory we compute now

$$\Gamma(m) = F(h) + \sum_i h_i m_i \tag{8.23}$$

and, after a little bit of elementary algebra, using identities such as

$$\arg\tanh m = \frac{1}{2} \log \frac{1+m}{1-m} \quad \text{and} \quad \log\cosh u = -\frac{1}{2} \log\left(1 - \tanh^2 u\right),$$

one finds

$$\Gamma^{\text{m.f.}} = -\frac{1}{2} \sum_{ij} J_{ij} m_i m_j + kT \sum_i \left[\frac{1+m_i}{2} \log \frac{1+m_i}{2} + \frac{1-m_i}{2} \log \frac{1+m_i}{2} \right]. \tag{8.24}$$

Note that this form of the free energy is very transparent: the first term of (8.24) is the interaction energy and the second term is entropy. Indeed, given a magnetization m, the probability that the spin is up is $(1 + m)/2$, and that it is down is $(1 - m)/2$. The entropy is $-k \sum_i p_i \log p_i$ and the r.h.s. of (8.24) is nothing but $E - TS$.

Exercise Infinite range model

One considers a system of N Ising spins, in which all pairs interact with a weak interaction of order $1/N$. *In an external field, the energy is*

$$\beta E(C) = -\frac{J}{2N} \sum_{i=1}^{N} \sum_{j=1}^{N} \sigma_i \sigma_j - h \sum_1^N \sigma_i.$$

Let us introduce the variable $\mu = \sum_1^N \sigma_i$ *and use the identity*

$$e^{\frac{J\mu^2}{2N}} = \sqrt{\frac{NJ}{2\pi}} \int_{-\infty}^{+\infty} dx e^{-\frac{NJx^2}{2} + J\mu x}.$$

1. *From the previous identity show that the partition function takes the form*

$$Z(h) = C \int dx e^{-N\phi(x,h)},$$

in which C is an h-independent constant.

2. *In the limit of a large volume, here large N, find the free energy per site* $f(h) = \lim_{N\to\infty} \frac{F(h)}{N}$.
3. *Determine the corresponding free energy* $\Gamma(m)$ *as a function of the magnetization and show that, for this model, the mean field theory is exact.*

Let us now show that the cumbersome expression (8.24) allows us to recover all the consequences of Landau theory in the vicinity of the critical point.

Homogeneous case

If the field h is uniform, $m_i = m$ is site independent and the free energy per site is given by

$$\gamma^{\text{m.f.}}(m) = -\frac{1}{2}\mathcal{J}m^2 + kT\left[\frac{1+m}{2}\log\frac{1+m}{2} + \frac{1-m}{2}\log\frac{1-m}{2}\right], \tag{8.25}$$

where $\mathcal{J}$ stands for

$$\mathcal{J} = \sum_j J_{ij} \tag{8.26}$$

(this last sum is independent of i by translation invariance). Expanding in powers of m this gives

$$\gamma^{\text{m.f.}}(m) = \frac{1}{2}a(T)m^2 + \frac{1}{4}bm^4 + O(m^6), \tag{8.27}$$

where

$$a = kT - \mathcal{J}, \qquad b = \frac{kT}{3}. \tag{8.28}$$

We are thus led back to the discussion that we had with Landau theory. The parameter a vanishes linearly and changes sign at a critical temperature

$$T_c = \frac{\mathcal{J}}{k}. \tag{8.29}$$

Since all the terms of degree higher than m^4 played no rôle in the dominant critical behaviour, the exact form (8.25) was only used to provide the coefficients a and b and nothing else.

Spatial correlations

Let us examine again the response of the magnetization at a point $\vec{r}_i$ to a change of the applied field at point $\vec{r}_j$ (the lattice sites are located at points labelled $i, j, \ldots$ but it is useful here to remember that the lattice point i is a vector $\vec{r}_i$ in d dimensions):

$$G_{ij} = \frac{\partial m_i}{\partial h_j}. \tag{8.30}$$

Here again, it is easier to calculate first its inverse

$$(G^{-1})_{ij} = \frac{\partial h_i}{\partial m_j} = \frac{\partial^2 \Gamma}{\partial m_i \partial m_j}. \tag{8.31}$$

In zero field, above T_c, we take two derivatives of (8.24) and set $m = 0$; this gives

$$(G^{-1})_{ij} = -J_{ij} + kT\delta_{ij}. \tag{8.32}$$

Translation invariance allows one to solve the required matrix inversion, to recover G_{ij}. Given a translation invariant matrix X_{ij}, i.e., $X_{ij} = X_{\vec{r}_i+\vec{u},\vec{r}_j+\vec{u}}$, in which $\vec{u}$ is any vector joining two lattice sites, one defines

$$x(\vec{q}) = \sum_j X_{ij}\, \mathrm{e}^{\mathrm{i}\vec{q}\cdot(\vec{r}_i-\vec{r}_j)}. \tag{8.33}$$

Let us take a simple 'cubic' lattice in d dimensions with lattice spacing a. Then this formula is inverted by integrating over $\vec{q}$ in a *Brillouin zone*, here in an interval for each component of $\vec{q}$ of size $2\pi/a$, for instance $(-\pi/a, \pi/a)$; then

$$X_{ij} = \frac{a^d}{(2\pi)^d}\int_{\mathrm{B.Z.}} \mathrm{d}^d q\; x(\vec{q})\, \mathrm{e}^{-\mathrm{i}\vec{q}\cdot(\vec{r}_i-\vec{r}_j)}. \tag{8.34}$$

Therefore, we define

$$\mathcal{J}(\vec{q}) = \sum_j J_{ij}\, \mathrm{e}^{\mathrm{i}\vec{q}\cdot(\vec{r}_i-\vec{r}_j)}, \tag{8.35}$$

$$\mathcal{G}(\vec{q}) = \sum_j G_{ij}\, \mathrm{e}^{\mathrm{i}\vec{q}\cdot(\vec{r}_i-\vec{r}_j)} \tag{8.36}$$

and

$$\mathcal{G}^{-1}(\vec{q}) = \sum_j (G^{-1})_{ij}\, \mathrm{e}^{\mathrm{i}\vec{q}\cdot(\vec{r}_i-\vec{r}_j)}, \tag{8.37}$$

then the Fourier transform of (8.32) gives

$$\mathcal{G}^{-1}(\vec{q}) = -\mathcal{J}(\vec{q}) + kT \tag{8.38}$$

and thus

$$\mathcal{G}(\vec{q}) = \frac{1}{-\mathcal{J}(\vec{q}) + kT}. \tag{8.39}$$

Let us note that $\mathcal{J}$ defined in (8.26) is given by

$$\mathcal{J} = \mathcal{J}(0) = kT_c \tag{8.40}$$

and thus we can write

$$\mathcal{G}(\vec{q}) = \frac{1}{\mathcal{J}(0) - \mathcal{J}(\vec{q}) + k(T - T_c)}. \tag{8.41}$$

This is general, but we can examine what happens for simple nearest-neighbour couplings:

$$\mathcal{J}(\vec{q}) = J(e^{iq_1 a} + e^{-iq_1 a} + \cdots + e^{iq_d a} + e^{-iq_d a}) \tag{8.42}$$

and

$$\mathcal{J}(0) - \mathcal{J}(\vec{q}) = 2J \sum_{l=1}^{d} (1 - \cos q_l a) \simeq Ja^2\vec{q}^2 + O(q^4). \tag{8.43}$$

Finally, we obtain

$$G_{ij} = \frac{a^d}{(2\pi)^d} \int_{\text{B.Z.}} d^d q \frac{e^{-i\vec{q}\cdot(\vec{r}_i - \vec{r}_j)}}{\mathcal{J}(0) - \mathcal{J}(\vec{q}) + k(T - T_c)}. \tag{8.44}$$

The behaviour for large distances, i.e., large $|\vec{r}_i - \vec{r}_j|$ is governed by the behaviour of the factor in front of the oscillating exponential at small q. If we assume an interaction such as $\mathcal{J}(0) - \mathcal{J}(\vec{q}) \sim \vec{q}^2$, as is the case for nearest-neighbour interactions, or any short-range interactions such that

$$\sum_j J_{ij}(\vec{r}_i - \vec{r}_j)^2 < \infty; \tag{8.45}$$

we are exactly in the situation studied with Landau theory.

8.3 A simpler alternative approach

We ended the previous section with the following conclusions: mean field theory is an approximation whose critical content is identical to Landau theory. We have thus to go beyond this approximation if we want to understand the quantitative failure of this approach. However, the spin model with which we started contains details that are superfluous, irrelevant. The only things that mattered in $\gamma(m)$ were the coefficients of m^2 and m^4; for the interaction all we need is that $\mathcal{J}(0) - \mathcal{J}(\vec{q}) \sim q^2$.

Let us consider the following alternative scheme. We start with microscopic degrees of freedom $\varphi(\vec{r})$ instead of the spins. Again, a symmetry group acts on φ, a simple Z_2 with $\varphi(\vec{r}) \to -\varphi(\vec{r})$, or a rotation group $\varphi_a(\vec{r}) \to R_{ab}\varphi_b(\vec{r})$, which leaves the energy invariant. The $\vec{r}$ are sites on a lattice. A configuration $\mathcal{C}$ consists of an assignment of values of $\varphi(\vec{r})$ at each lattice site. Each component of $\varphi(\vec{r})$ varies from $-\infty$ to $+\infty$. The probability distribution for $\varphi(\vec{r})$ is chosen to be

$$p(\varphi(\vec{r})) = N \, \mathrm{e}^{-c\varphi(\vec{r})^2 - d\varphi(\vec{r})^4}, \tag{8.46}$$

in which c, d are constants and N is a normalization. Nearest neighbours are coupled and the coupling energy is

$$E(\varphi(\vec{r})) = J \sum_{\vec{r}} \sum_{\alpha=1}^{d} (\varphi(r_\alpha + a e_\alpha) - \varphi(r_\alpha))^2, \tag{8.47}$$

in which e_α is the unit vector in the direction α. If we expand the square in (8.47) we obtain φ^2 terms and a product at two neighbouring sites. Again, this interaction favours slow spatial variations of $\varphi(\vec{r})$. Instead of (8.47), it is often convenient to replace

$$\sum_{\alpha=1}^{d} (\varphi(r_\alpha + a e_\alpha) - \varphi(r_\alpha))^2 \to a^2 (\nabla\varphi)^2 \tag{8.48}$$

and similarly

$$\sum_{\vec{r}} \to a^{-d} \int \mathrm{d}^d r \tag{8.49}$$

but we have to remember that, in case of difficulties, the continuum model is 'regularized' by the underlying lattice.

If we combine the probability (8.46) and the energy (8.47), we end up with a model with an exponential weight e^{-A} with

$$A\{\varphi\} = \int \mathrm{d}^d r \left\{ \frac{1}{2}(\nabla\varphi)^2 + \frac{1}{2} r(T)\varphi^2 + \frac{\lambda}{4}\varphi^4 \right\}. \tag{8.50}$$

We have rescaled φ so that the coefficient of the gradient of φ is $\frac{1}{2}$; hence we have absorbed the inverse temperature β in this rescaling, but we should remember that the coefficient of φ^2 is temperature dependent and may change sign.

Therefore, the partition function of this model, if we add a spatially varying magnetic field $\mathbf{h}(\vec{r})$, is

$$Z\{h\} = \int \prod_{\vec{r}} \mathrm{d}\varphi(\vec{r}) \, \mathrm{e}^{-A\{\varphi\} + \int \mathrm{d}^d r \mathbf{h}(\vec{r}) \cdot \varphi(\vec{r})}. \tag{8.51}$$

Let us check that the mean field and Landau theories correspond to the simplest approximations to (13.3), namely we first look for the configuration of $\varphi(\vec{r})$ with the largest weight, i.e., which minimizes the combination $A - \int h \cdot \varphi$. This configuration $\hat{\varphi}$ is thus the solution of

$$\frac{\delta A}{\delta\varphi(\vec{r})} - \mathbf{h}(\vec{\mathbf{r}}) = 0. \tag{8.52}$$

At lowest order we keep only this configuration in the partition function, neglecting all fluctuations of φ around $\hat{\varphi}$. This gives

$$\log Z^{(0)}\{h\} = -A\{\hat{\varphi}\} + \int \mathrm{d}^d r\, \mathbf{h}(\vec{r}) \cdot \hat{\varphi}(\vec{r}). \tag{8.53}$$

We want to calculate the free energy $\Gamma\{m\}$ and we follow the usual steps:

1. Find the magnetization:

$$\mathbf{m}(\vec{r}) = -\frac{\delta F^{(0)}}{\delta h(\vec{r})} = \frac{\delta \log Z^{(0)}}{\delta h(\vec{r})} \tag{8.54}$$

(we have absorbed β in the rescaling of φ). In performing this derivative one should remember that $\hat{\varphi}$ depends upon h; however, the relation (8.52) that defines $\hat{\varphi}$ leads to the simple result

$$\mathbf{m}(\vec{r}) = \hat{\varphi}(\vec{r}) \tag{8.55}$$

(indeed

$$\frac{\delta \log Z^{(0)}}{\delta h(\vec{r})} = \hat{\varphi}(\vec{r}) + \int \mathrm{d}^d s\, \mathbf{h}(\vec{s}) \cdot \frac{\delta\varphi(\vec{s})}{\delta h(\vec{r})} - \int \mathrm{d}^d s \frac{\delta A}{\delta\varphi(\vec{s})} \frac{\delta\varphi(\vec{s})}{\delta h(\vec{r})}$$

but the last two terms cancel from (8.52)).

2. Apply the Legendre transform

$$\Gamma^{(0)}\{m\} = F^{(0)}\{h\} + \int \mathrm{d}^d r\, \mathbf{h}(\vec{r}) \cdot \mathbf{m}(\vec{r}), \tag{8.56}$$

which reduces here simply to

$$\Gamma^{(0)}\{m\} = A\{m\} = \int \mathrm{d}^d r \left\{ \frac{1}{2}(\nabla\mathbf{m})^2 + \frac{1}{2} r(T)\mathbf{m}^2 + \frac{\lambda}{4}\mathbf{m}^4 \right\}, \tag{8.57}$$

when we take into account the expression $F^{(0)} = -\log Z^{(0)}$ and (8.55).

We have thus reached the goal that we had in mind: we have found a simpler model in which the lowest approximation, in which we neglect all fluctuations of the order parameter around its most likely value, reduces to the Landau theory. Up to now, nothing new then, but we are in a position to study whether the approximation of order zero, that we have made, makes sense. If we took into account the small fluctuations about $\hat{\varphi}$ would it affect the previous analysis or not? This will be examined in the following chapter.

9

Beyond the mean field theory

9.1 The first correction to the mean-field free energy

In the previous chapter we have limited ourselves to the most likely value $\hat{\varphi}$ of the 'spin' and neglected all fluctuations. We are now in a position to study whether this approximation was legitimate. We thus consider 'small' deviations from $\hat{\varphi}$ and write

$$\varphi = \hat{\varphi} + \psi, \tag{9.1}$$

and expand the weight $A - h\varphi$ to second order in ψ (the first order in ψ vanishes by definition of $\hat{\varphi}$):

$$\begin{aligned} A\{\varphi\} &- \int \mathrm{d}^d r \, \mathbf{h}(\vec{r}) \cdot \varphi(\vec{r}) \\ &= A\{\hat{\varphi}\} - \int \mathrm{d}^d r \, \mathbf{h}(\vec{r}) \cdot \hat{\varphi}(\vec{r}) \\ &\quad + \frac{1}{2} \int \mathrm{d}^d r \, \mathrm{d}^d s \, \psi(\vec{r}) \left. \frac{\delta^2 A}{\delta\varphi(\vec{r})\delta\varphi(\vec{s})} \right|_{\hat{\varphi}} \psi(\vec{s}) + O(\psi^3). \end{aligned} \tag{9.2}$$

This second derivative of A is given here by the kernel

$$K(\vec{r}, \vec{s}) = \left. \frac{\delta^2 A}{\delta\varphi(\vec{r})\delta\varphi(\vec{s})} \right|_{\hat{\varphi}} = \delta(\vec{r} - \vec{s})(r + 3\lambda\hat{\varphi}^2 - \vec{\nabla}^2). \tag{9.3}$$

(For simplicity, we limit ourselves in this chapter to a one-component φ; for an n-component field φ_α one obtains a kernel $K_{\alpha\beta}(\vec{r}, \vec{s})$. It is recommended that the interested reader repeat the calculations of this chapter with an n-component order parameter.) Therefore, taking into account the Gaussian fluctuations over $\hat{\varphi}$, we have

$$\mathrm{Z}^{(1)} = \mathrm{Z}^{(0)} \int \prod_{\vec{r}} \mathrm{d}\psi(\vec{r}) \mathrm{e}^{-\frac{1}{2} \int \mathrm{d}^d r \mathrm{d}^d s \, \psi(\vec{r}) K(\vec{r}, \vec{s}) \psi(\vec{s})}. \tag{9.4}$$

This Gaussian integral over ψ gives (up to an irrelevant power of (2π))

$$Z^{(1)} = Z^{(0)} \frac{1}{\sqrt{\det K}} \tag{9.5}$$

i.e., the free energy

$$F^{(1)} = F^{(0)} + \frac{1}{2} \mathrm{Tr} \log K \tag{9.6}$$

(we have used the identity $\log \det K = \mathrm{Tr} \log K$).

We have thus to find the eigenvalues of K. One cannot say much in general, but if the external field is constant in space, the same is true of $\hat{\varphi}$ and then the kernel $K(\vec{r}, \vec{s}) = K(\vec{r} - \vec{s})$ is invariant by translation and a Fourier transformation will give the eigenvalues of K.

It remains to calculate the Legendre transformation to obtain $\Gamma^{(1)}$. The result is simple, but one has to take care. To simplify notation, let us write

$$F^{(1)} = F^{(0)} + \Delta,$$

in which Δ stands for the correction $\frac{1}{2}\mathrm{Tr} \log K$. Then

$$m^{(1)} = m^{(0)} - \frac{\partial \Delta}{\partial h} \tag{9.7}$$

$$\begin{aligned}
\Gamma^{(1)}\{m^{(1)}\} &= F^{(0)} + \Delta + h m^{(1)} \\
&= F^{(0)} + h m^{(0)} + \Delta - h \frac{\partial \Delta}{\partial h} \\
&= \Gamma^{(0)}\{m^{(0)}\} + \Delta - h \frac{\partial \Delta}{\partial h} \\
&= \Gamma^{(0)}\left\{m^{(1)} + \frac{\partial \Delta}{\partial h}\right\} + \Delta - h \frac{\partial \Delta}{\partial h} \\
&= \Gamma^{(0)}\{m^{(1)}\} + h \frac{\partial \Delta}{\partial h} + \Delta - h \frac{\partial \Delta}{\partial h} \\
&= \Gamma^{(0)}\{m^{(1)}\} + \Delta.
\end{aligned} \tag{9.8}$$

We end up with the alleged correction to the Landau theory:

$$\Gamma^{(1)}\{m\} = \int \mathrm{d}^d r \left\{ \frac{1}{2} (\nabla \mathbf{m})^2 + \frac{1}{2} r(T) m^2 + \frac{\lambda}{4} m^4 \right\} + \frac{1}{2} \mathrm{Tr} \log K, \tag{9.9}$$

with

$$K(\vec{r}, \vec{s}) = \delta(\vec{r} - \vec{s})(r + 3\lambda m^2 - \vec{\nabla}^2). \tag{9.10}$$

Remark

The expression (9.9) is called the one-loop correction to the fluctuationless mean-field approximation. The reason is that this free energy is the generating functional of all diagrams with one loop of integration. Clearly, the procedure that we have used here can be pushed to higher orders, but we will understand the main features of the problem from this one-loop functional.

For arbitrary $m(\vec{r})$ we cannot go further, but if the magnetization is uniform, $m(\vec{r}) = m$, the problem becomes simple since the kernel K becomes translation invariant. Let us consider a simple box of volume $V = L^d$ with periodic boundary conditions. The normalized function

$$u_{\vec{q}}(\vec{r}) = \frac{1}{\sqrt{V}} e^{i\vec{q}\cdot\vec{r}}, \tag{9.11}$$

in which all the d components of q are integer multiples of $2\pi/L$, are periodic, and eigenfunctions of K since

$$\int d^d s \delta(\vec{r} - \vec{s})(r + 3\lambda m^2 - \vec{\nabla}^2) u_{\vec{q}}(\vec{s}) = (r + 3\lambda m^2 + \vec{q}^2) u_{\vec{q}}(\vec{r}). \tag{9.12}$$

Thereforc,

$$\mathrm{Tr}\log K = \sum_{\vec{q}} \log(r + 3\lambda m^2 + \vec{q}^2), \tag{9.13}$$

in which the $\vec{q}$ have quantized components $2\pi n/L$. In the large volume limit, these sums converge to the integral

$$\mathrm{Tr}\log K = \frac{V}{(2\pi)^d} \int d^d q \log(r + 3\lambda m^2 + \vec{q}^2). \tag{9.14}$$

Then the free energy per unit volume is given by

$$\gamma^{(1)}(m) = \frac{1}{2} r(T) m^2 + \frac{\lambda}{4} m^4 + \frac{1}{2} \int \frac{d^d q}{(2\pi)^d} \log\left(\frac{r + 3\lambda m^2 + \vec{q}^2}{r + \vec{q}^2}\right), \tag{9.15}$$

in which we have subtracted an m-independent constant so that $\gamma^{(1)}(0) = 0$. We should remember at this stage that the underlying lattice has the effect of limiting the integral for large $|\vec{q}|$ by a cut-off given by the size of the Brillouin zone, which is proportional to the inverse of the lattice spacing a.

9.2 Physical consequences

There are several physical observables that we might draw from (9.15); they would all lead to a similar analysis. Let us examine only one, for instance the zero-field magnetic susceptibility above T_c:

$$\chi = \left.\frac{\partial m}{\partial h}\right|_{h=0}, \tag{9.16}$$

or rather,

$$1/\chi = \left.\frac{\partial^2 \gamma}{\partial m^2}\right|_{m=0} = r(T) + 3\lambda \int \frac{\mathrm{d}^d q}{(2\pi)^d} \frac{1}{q^2 + r}. \tag{9.17}$$

In mean field theory, the susceptibility diverges at the critical temperature given by $r = 0$. We see that the fluctuations have shifted the critical temperature since r_c is no longer zero but it is given by

$$0 = r(T_c) + \frac{3\lambda}{2} \int \frac{\mathrm{d}^d q}{(2\pi)^d} \frac{1}{q^2 + r_c}. \tag{9.18}$$

Subtracting (9.18) from (9.17) we end up with

$$1/\chi = [r(T) - r(T_c)] \left(1 - \frac{3\lambda}{2} \int \frac{\mathrm{d}^d q}{(2\pi)^d} \frac{1}{(q^2 + r)(q^2 + r_c)}\right). \tag{9.19}$$

In the last integral we can replace r_c, which is of order λ, by zero since it would yield a contribution of higher order, and for the same reason r by $r(T) - r(T_c)$, which is proportional to $t = (T - T_c)/T_c$. Taking the constant of proportionality equal to one, we end up with

$$1/\chi = t\left[1 - \frac{3\lambda}{2} \int \frac{\mathrm{d}^d q}{(2\pi)^d} \frac{1}{(q^2 + t)q^2}\right]. \tag{9.20}$$

At this stage, it is important to understand the logic of this calculation. We have expanded in perturbation theory, to first order, the fluctuations of the order parameter about its most likely value, to go beyond the Landau theory. If the correction is small, the Landau theory will be justified; if it is large, the Landau theory breaks down, but so does this calculation as well: if it turns out that the first alleged correction is larger than the main term, this expansion is certainly not to be trusted. The next term would presumably be even larger and so on.

Returning to (9.20) we see that the first term gives $\chi \sim 1/t$. If the integral in the bracket remains finite when $t \to 0$ then this result will remain true in our calculation as well. So everything relies on the integral

$$I(t) = \int \frac{\mathrm{d}^d q}{(2\pi)^d} \frac{1}{(q^2 + t)q^2}. \tag{9.21}$$

1. If $I(0)$ is finite, χ still diverges as $1/t$. But

$$I(0) = \int \frac{\mathrm{d}^d q}{(2\pi)^d} \frac{1}{(q^2)^2} \tag{9.22}$$

   and we have no problem for large q since there is a lattice cut-off, but the integral blows up at small q unless $d > 4$. Therefore, within this approach, the Landau theory remains valid above four dimensions; the critical temperature is modified by the fluctuations, the coefficient of $1/t$ also, but not the nature of the singularity.

2. However, $I(t)$ diverges if $d \leq 4$; a logarithmic divergence in four dimensions, a power $t^{-(4-d)/2}$ below four dimensions. The Landau theory breaks down but perturbation theory as we have described it is useless since the first correction goes to infinity, and the next ones even more.

Several remarks are now in order:

- For $d > 4$ the mean-field singularity is not modified by the fluctuations. This was shown here at first order, but the analysis can be extended to arbitrary order. This result, in itself, appeared initially as very surprising. For a long time it was believed that the mean-field picture would be more and more exact when the dimension of space increased: indeed the number of neighbours of a given site increases with d and then it is expected that the fluctuations of the mean field decrease with d. In this line of reasoning there were attempts in the past to start with the mean field for infinite d and expand the critical exponents in powers of $1/d$. This is not what happens: if the exponents change with d for $d < 4$, they remain constantly equal to their mean-field values for $d > 4$. (I shall show below that at $d = 4$ there are logarithmic corrections to the mean-field picture.)

  Can one find a more intuitive understanding of the rôle of four dimensions? The following heuristic argument, from J. des Cloizeaux, is a step in that direction. It applies to self-avoiding walks, the $n = 0$ limit of our problem. Consider two self-avoiding walks; the two chains have a fractal dimension $d_{\mathrm{f}} = 1/\nu$. In dimension d, two manifolds of dimensions d_1 and d_2 intersect 'generically' only if $d_1 + d_2 \geq d$ (like two planes, or a plane and a line in three dimensions). For the two self-avoiding walks, the excluded-volume rule of non-intersection will thus matter only if $2d_{\mathrm{f}} \geq d$. On the other hand, if $2d_{\mathrm{f}} < d$, one expects the excluded-volume rule to be irrelevant, since the probability of intersection is zero, and thus the chains should be Brownian. Then one has $\nu = 1/2$, $d_{\mathrm{f}} = 2$: hence, if $2d_{\mathrm{f}} = 4 < d$ the exponent of the self-avoiding walk coincides with that of the simple Brownian random walk, the equivalent here of the mean-field value.
- Returning to equation (9.20) one sees that, if the mean-field behaviour breaks down close to T_{c} in fewer than four dimensions, further from T_{c} fluctuations give only a small contribution, proportional to $\lambda/t^{2-d/2}$. This parameter, introduced by Ginzburg, fixes the size of the critical region: as long as it remains small, one sees a mean-field behaviour. When it becomes large, mean field theory breaks down.

Conclusion Understanding the critical behaviour in dimensions $d \leq 4$ is a non-perturbative problem, i.e., a priori, a difficult problem, which remained unsolved until Wilson's work on the renormalization group (1971)[1] enlightened and really solved the problem.

[1] K. G. Wilson, *Phys. Rev. B*, **4** (1971) 3174 and 3184.

10

Introduction to the renormalization group

The initial formulation of the renormalization group appeared in the study of quantum electrodynamics (QED) in the high-energy limit where the momenta of the particles (electrons, positrons, photons) were much larger than the rest mass of the electron (Gell-Mann and Low, 1954).[1] It is then natural to neglect the electron mass from the beginning (rather than calculating Feynman diagrams and taking their high-energy limit). However, the theory with massless electrons has to be handled with some care: the usual definition of the renormalized charge, normally defined as the electron–photon vertex in the limit of vanishing momenta, leads to an infrared divergence. Then the definition of the renormalized charge has to be modified as being the value of the same vertex but at some arbitrary non-zero momenta. This introduces a new length scale to the theory (the inverse of an arbitrary momentum in units $\hbar = c = 1$). Physical quantities should not depend on this arbitrariness, and different charges of the electron, related to different scales of definition, lead to the same physics.

It was only in the beginning of the 1970s that the study of dilatation invariance in field theory, extended by K. Wilson to the vicinity of the critical point and similar physical problems, threw light on the true meaning of the renormalization group.[2] In a classical (non-quantum) theory, such as massless electrodynamics, the charge of the electron is dimensionless in dimension (3+1), and the Maxwell–Dirac equations are scale invariant, i.e., unchanged under

$$x \to \lambda x, \quad A_\mu(x) \to \lambda^{-1} A_\mu(\lambda x), \quad \psi(x) \to \lambda^{-3/2}\psi(\lambda x). \tag{10.1}$$

However, quantum theory breaks this dilatation invariance. Indeed, it is necessary to introduce an extra length to regularize the theory, such as the inverse of

[1] M. Gell-Mann and F. E. Wilson, *Phys. Rev.*, **95** (1954) 1300.

[2] K. G. Wilson, *Phys. Rev. B*, **4** (1971) 3174 and 3184; see also *Phys. Rev. D*, **2** (1970) 1478. C. G. Callan, *Phys. Rev. D*, **2** (1970) 1541. K. Symanzik, *Commun. Math. Phys.*, **18** (1970) 227.

a momentum cut-off, or massive regularizing fields (Pauli–Villars), or dimensional regularization, in which the charge of the electron acquires the dimension of a length raised to the power $\frac{1}{2}(4-d)$. Therefore, quantum theory is not scale invariant at any length scale, but only asymptotically at large distances with respect to the small length (or inverse of a large momentum) that we had to introduce.

10.1 Renormalized theories and critical points

The vicinity of the critical point is, as we saw in Chapter 7, scale invariant: this manifests itself when the correlation length ξ is very large compared with the lattice spacing a. The analogy between criticality and renormalization is manifest. In its Euclidean (imaginary time) formulation the weight of the path integral is analogous to a Boltzmann factor. The regularization introduces a length scale, for instance $a \sim \Lambda^{-1}$, the inverse of the momentum cut-off. The physics is related to situations in which Λ is much greater than the physical masses and momenta, i.e., length scales much larger than a. One must keep in mind that within this regularization the gradient of a field represents the difference of its values at points that are at distance a apart. For a simple scalar Euclidean action

$$A = \int \mathrm{d}x((\partial_\mu \varphi)^2 + V(\varphi)), \tag{10.2}$$

the interaction term V is local; it does not couple distant points. The gradient term couples points at distance a. One would expect, therefore, that the range of spatial correlations would be not much larger than a: this would mean that the physical mass, which is the inverse of this range, would be of the order of the cut-off. It is thus necessary to adjust the parameters in an appropriate way to create long-range correlations, and physical masses much smaller than the cut-off. In statistical mechanics, this requires adjusting the temperature to be near T_c and the external field to be very small. In field theory, the existence of a renormalized theory, in which the cut-off is much larger than the masses, implies appropriately fixed bare parameters of the action (bare masses, bare coupling constants).

10.2 Kadanoff block spins

In 1966, Kadanoff[3] introduced a very simple picture to describe an interaction that would vary with the scale. This picture is very approximate, it is much too rudimentary to be able to lead to realistic calculations, but it is simple and thus useful to have in mind when we go to more elaborate methods.

[3] L. P. Kadanoff, *Physics*, **2** (1966) 263.

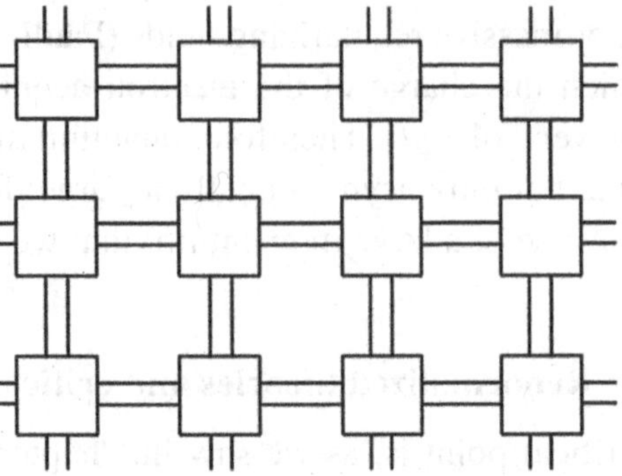

Figure 10.1 Kadanoff blocks

Let us consider a two-dimensional Ising model on a square lattice, lattice spacing a, with

$$\beta E(\mathcal{C}) = -J \sum_{\langle ij \rangle} \sigma_i \sigma_j \tag{10.3}$$

(a factor β has been included in J). We now divide the lattice into blocks of four spins (Fig. 10.1).

These blocks form themselves into a lattice of spacing $2a$. For each block of four spins, we define a block spin

$$S_b = \frac{1}{4}(\sigma_1 + \sigma_2 + \sigma_3 + \sigma_4), \tag{10.4}$$

if $(\sigma_1, \sigma_2, \sigma_3, \sigma_4)$ are the four vertices of the block b. Near T_c, where the range of correlation is very large, the spatial fluctuations of small wavelength that would destroy long-range order, are rare. Therefore, out of the 16 configurations for the block, we retain only the two in which the four spins are alike, either all up, or all down. Then S_b takes only the values ± 1. The constant $(2J)$ of the initial interaction (10.3) is the difference of energy between two nearby spins parallel or antiparallel. Let us term this difference of energy for the two blocks J', which depends on J:

$$J' = f(J) \tag{10.5}$$

and let us leave the function f unspecified at this stage. If $\xi(J)$ is the correlation length measured as the number of steps on the initial lattice, on the block lattice it is divided by 2 (a becomes $2a$). Therefore,

$$\xi(J') = \frac{1}{2}\xi(J). \tag{10.6}$$

At the critical point J_c, $\xi(J_c)$ is infinite; therefore, for the corresponding $J' = f(J_c)$, ξ is also infinite. Therefore, the critical point is a fixed point of (10.5), namely a solution of:

$$J_c = f(J_c). \tag{10.7}$$

Near J_c, one can thus linearize the relation (10.5)

$$J_c - J' \simeq f'(J_c)(J_c - J). \tag{10.8}$$

If the divergence of ξ near J_c is governed by an exponent ν

$$\xi(J) \sim (J_c - J)^{-\nu}, \tag{10.9}$$

then we must have

$$(J_c - J')^{-\nu} = \frac{1}{2}(J_c - J')^{-\nu}, \tag{10.10}$$

which gives

$$[f'(J_c)]^{-\nu} = \frac{1}{2} \quad \text{or} \quad \nu = \frac{\log 2}{\log f'(J_c)}. \tag{10.11}$$

This very primitive method, although not really satisfactory as argued below, nevertheless contains in essence the features that we will find in all the renormalization group methods:

1. A change of the elementary scale induces a transformation of the parameters characterizing the interactions.
2. A critical temperature as a fixed point of this transformation.
3. Critical exponents related to the linearized transformation near the fixed point.

In order to complete this programme, it remains to compute $f(J)$. The naive estimate given by the difference of energy between two parallel adjacent blocks and two antiparallel would lead to $2J' = 4J$, i.e., no non-zero fixed point. So the method remained as an interesting idea but not successful until it was reanalyzed by K. Wilson in 1971.

10.3 Examples of real space renormalization groups: 'decimation'

One dimension

This is a trivial and uninteresting problem, since the correlation length remains finite at any non-zero temperature, but it is easy to treat. In Chapter 2, we have computed the correlation length and found

$$\xi = \frac{1}{\log \coth J} \tag{10.12}$$

(we have replaced βJ with J).

To reduce the number of degrees of freedom, we first treat the odd sites; the spin σ_3 is coupled to σ_2 and σ_4. If we first sum over σ_3, we have

$$\sum_{\sigma_3=\pm 1} \mathrm{e}^{J\sigma_3(\sigma_2+\sigma_4)} = A\mathrm{e}^{J'\sigma_2\sigma_4}. \tag{10.13}$$

It is easy to check that this is an identity with

$$\mathrm{e}^{2J'} = \cosh 2J \qquad A = (4\cosh 2J)^{1/2}. \tag{10.14}$$

This identity may be used repeatedly to sum over all the spins at odd lattice sites. (Let us assume that the number of lattice sites is even; it clearly doesn't matter in the thermodynamic limit.) We are thus led to an Ising model for the remaining even sites with $a \to 2a$ and $J \to J'$.

The first relation (10.14) is easy to transform into

$$\tanh J' = \tanh^2 J \tag{10.15}$$

and it is easy to check with the explicit relation (10.12) that, since the remaining even spins form a lattice with $a \to 2a$, ξ has been divided by two:

$$\xi(J') = \frac{1}{2}\xi(J). \tag{10.16}$$

Finally, we have shown that

$$Z_N(J) = (4\cosh 2J)^{N/2} Z_{N/2}(J') \tag{10.17}$$

or, for the free energy per site, $f = \log Z/N$ to

$$f(J) = \frac{1}{2}f(J') - \frac{1}{2}\log 4\cosh 2J, \tag{10.18}$$

from which one can easily recover the exact free energy. But let us return to (10.15) to understand its content.

The effect of successive iterations is to divide the correlation length by two each time. Therefore, for any initial value of J the iteration moves away from the fixed point at $J = \infty$ (zero-temperature) to converge to the $J = 0$ (infinite temperature) fixed point, at which the correlation length vanishes (Fig. 10.2). Therefore, one discovers in this language that at any temperature the system is disordered.

More than one dimension

One can try the same strategy of summing over half the spins in any dimension. For two dimensions, one sums over the lattice points surrounded by squares on the

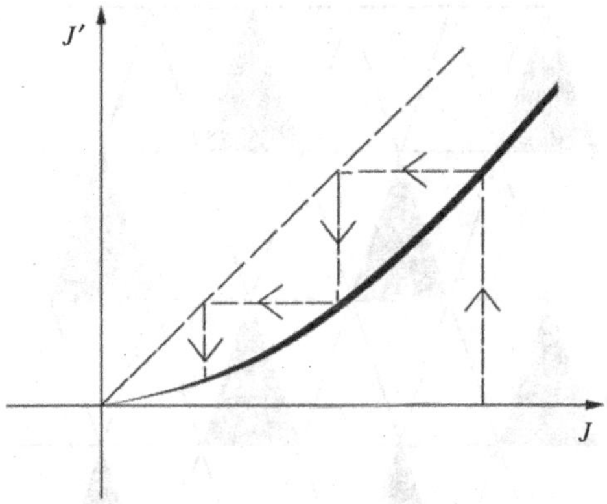

Figure 10.2 Iteration for a one-dimensional decimation

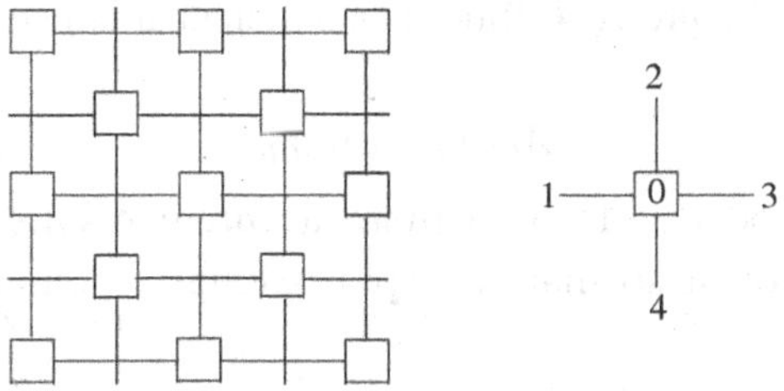

Figure 10.3 A two-dimensional decimation

figure, creating a new lattice of the remaining sites with $a \to a\sqrt{2}$. A given spin σ_0 is coupled to four neighbours (Fig. 10.3) but

$$\sum_{\sigma_0=\pm 1} \mathrm{e}^{J\sigma_0(\sigma_1+\sigma_2+\sigma_3+\sigma_4)} \neq A\mathrm{e}^{J'(\sigma_1\sigma_2+\sigma_2\sigma_3+\sigma_3\sigma_4+\sigma_4\sigma_1)}, \tag{10.19}$$

and

$$\sum_{\sigma_0=\pm 1} \mathrm{e}^{J\sigma_0(\sigma_1+\sigma_2+\sigma_3+\sigma_4)} = A\mathrm{e}^{J'(\sigma_1\sigma_2+\sigma_2\sigma_3+\sigma_3\sigma_4+\sigma_4\sigma_1)}\,\mathrm{e}^{K'(\sigma_1\sigma_3+\sigma_2\sigma_4)+L'(\sigma_1\sigma_2\sigma_3\sigma_4)}. \tag{10.20}$$

The decimation generates a next-nearest-neighbour interaction and a four-spin interaction, which were not present in the initial model. One could then say that the initial model was too special and that one should start with a model that, from the beginning, contains the three coupling constants J, K, L, but then the same decimation not only is no longer possible exactly, but generates even more couplings. Therefore, real space renormalization group transformations of the sort considered here are more an art than a science. Later, we shall see why a small number of interaction parameters should suffice, but it is not very clear at this stage.

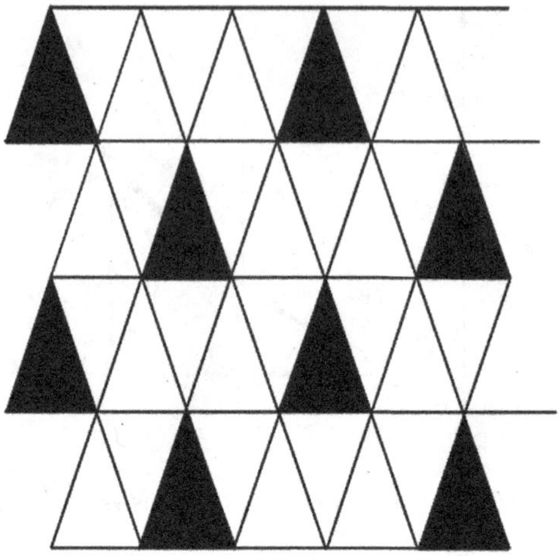

Figure 10.4 Blocks on a triangular lattice

Another example

Sometimes the method can lead to more accurate results. Consider a triangular lattice and partition it so that any spin belongs to one triangular block only (Fig. 10.4).

The block spins (the black triangles) form a triangular lattice, in which $a \to a\sqrt{3}$. Define a block spin as

$$\sigma_b = \text{sign}(\sigma_1 + \sigma_2 + \sigma_3), \tag{10.21}$$

a simple majority rule. One then defines

$$\mathrm{e}^{-E\{\sigma_b\}} = \sum_{\sigma,\ \text{with } \sigma_b \text{ fixed}} \mathrm{e}^{-E\{\sigma\}}. \tag{10.22}$$

If this calculation led to an Ising model of the block spins with nearest-neighbour couplings the problem would be solved. However, it does yield, like the decimation, new couplings, and the calculations can only be done approximately.

Percolation

This model has features similar to that of critical points, as was shown in Chapter 6 by connecting it to a Potts model. One starts with an infinite periodic lattice; each site may be occupied with probability p, or empty with probability $(1 - p)$.

A cluster is a set of occupied sites connected by lattice bounds. At low p the clusters are small, but one can prove that beyond a threshold p_c there is an infinite cluster with probability one. For p below p_c the diameter ξ of the largest cluster diverges when p becomes close to p_c as $(p_\mathrm{c} - p)^{-\nu}$. The following approximate renormalization group does a good job. Start with the triangular lattice of the previous figure, and partition it into blocks (the black triangles of the figure).

We will decide that a block is occupied or empty by the majority rule: if at least two of the three sites are occupied, the block is declared occupied. (This is only approximate for blocks across the boundary of the cluster: they can be declared occupied even with one site out; conversely they can be empty even with one site in the cluster.) Our assumptions lead to

$$p' = p^3 + 3p^2(1-p). \tag{10.23}$$

The block triangles form themselves into a triangular lattice with spacing $a \to a\sqrt{3}$:

$$\xi(p') = \frac{\xi(p)}{\sqrt{3}}. \tag{10.24}$$

At a fixed point ξ diverges, and there are three fixed points $p = 0, \frac{1}{2}, 1$. Clearly, an empty or a full lattice are not what we are studying, so $p_c = \frac{1}{2}$ (which happens to be exact). A linearization of (10.23) near p_c gives

$$p_c - p' \simeq \frac{3}{2}(p_c - p) \tag{10.25}$$

and thus

$$\left(\frac{3}{2}\right)^{-\nu} = \frac{1}{\sqrt{3}} \tag{10.26}$$

i.e., $\nu \simeq 1.35$, whereas the exact result (this problem has been solved in two dimensions) is $\nu = \frac{4}{3}$. So, at last, this is an example in which real space renormalization grouping is reasonable.

Exercises

Migdal–Kadanoff approximation (see Fig. 10.5)

(A) Dimension one

One considers a linear chain of Ising spins with nearest-neighbour interactions, lattice spacing a,

$$\beta E(C) = -J \sum_{\langle ij \rangle} \sigma_i \sigma_j.$$

One sums over the $(n-1)$ spins, between two sites at distance na on the lattice (the decimation above being the simplest $n = 2$ case).

1. *Prove that the remaining spins are again nearest-neighbour Ising spins (on the new lattice) with*

$$\tanh K' = (\tanh K)^n.$$

2. *Deduce from the above a functional relation for the correlation length $\xi(K)$; find the solution to this functional equation (take as variable $x = \tanh K$).*

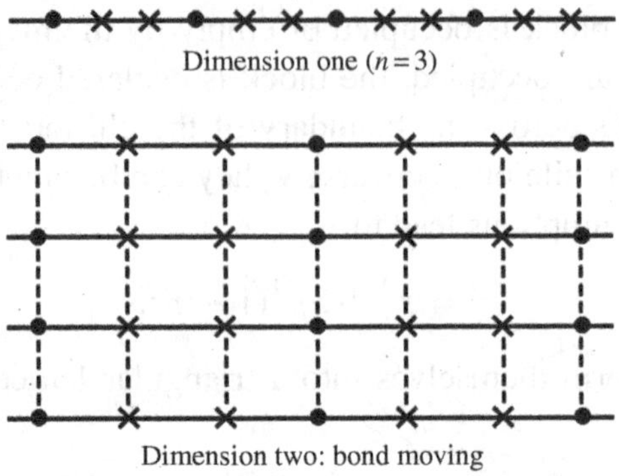

Figure 10.5 Bond moving approximation

(B) Two dimensions

One now considers a square lattice in dimension two and assumes that the interaction constants K_x and K_y in the horizontal and vertical directions are unequal. A sequence of two transformations will allow us to go from a lattice with lattice constant a, to a lattice with constant na.

(i) *First transformation: vertical bond moving*
As a simplifying approximation, one moves the bonds as shown in Fig. 10.5. The resulting sites marked by a cross have only horizontal couplings; however, the circled sites have a vertical coupling nK_y and horizontal couplings unchanged. Clearly, the ground state energy is unchanged.

Sum over the crossed sites and show that the circled sites make a nearest-neighbour Ising model with constants $\tilde{K}_x$ and $\tilde{K}_y$.

(ii) *Second transformation: horizontal bond moving*
After the first transformation, the lattice had elementary cells $a \times na$. To recover a square cell, one repeats the previous bond moving after exchange of the x and y axes.

1. *Show that after these two transformations one recovers an Ising model with constants K'_x and K'_y related to K_x and K_y.*
2. *Show on a graph that the relation $K_x \to K'_x$ possesses fixed points and identify the critical one.*
3. *The procedure is in fact non-symmetric with respect to the two directions: even if $K_x = K_y$, it yields $K'_x \neq K'_y$. To restore the symmetry, one continues the relations to non-integer values of n and studies the limit in which n approaches one. Writing*

$$n = 1 + \epsilon$$

one has

$$K'_x = K_x + \epsilon \Delta_x + O(\epsilon^2) \quad \text{and} \quad K'_y = K_y + \epsilon \Delta_y + O(\epsilon^2).$$

Determine Δ_x and Δ_y in terms of K_x and K_y and verify that if $K_x = K_y$ the (x, y) symmetry is restored.

4. *If one restricts the problem to $K_x = K_y$, show that the exact critical temperature of the two-dimensional Ising model is a fixed point of the relations found in the previous question.*
5. *Find from these same relations an approximate value for the correlation length exponent ν.*

10.4 Structure of the renormalization group equations

Before we return to field theory with explicit methods for deriving the renormalization group equations, it is worth examining abstractly the consequences of these scale transformations. Let us imagine that, through a decimation, block spin, or any similar method, one multiplies the basic length scale a by a factor $l > 1$: $a \to la$; the number of sites in d dimensions is divided by N, giving N/l^d. The constants of interaction that we had introduced are denoted $J_1, \ldots, J_p$ (neighbours, next-nearest neighbours, four-spin interactions, etc.). Assume that we know how these parameters transform under the rescaling

$$a \to la,\ \xi \to \xi/a,\ N \to N/l^d,\ J_\alpha \to J'_\alpha = f_\alpha(J_1, \cdots, J_p). \tag{10.27}$$

The correlation length is infinite at a fixed point $J^\star_\alpha$ of this transformation, and we assume again that the transformation is not singular near the fixed point. In its vicinity one may thus expand and write the linearized relation

$$J'_\alpha - J_\alpha = \sum_\beta M_{\alpha\beta}(l)(J_\beta - J^\star_\beta), \tag{10.28}$$

in which

$$M_{\alpha\beta}(l) = \left.\frac{\partial f_\alpha}{\partial J_\beta}\right|_{J^\star}. \tag{10.29}$$

This matrix is not necessarily symmetric but we assume that it has a spectrum

$$\sum_{\beta=1}^{p} x^{(i)}_\beta M_{\beta\alpha} = \lambda^{(i)} x^{(i)}_\alpha. \tag{10.30}$$

The linear combination of interaction terms

$$K^{(i)} = \sum_{\alpha=1}^{p} x^{(i)}_\alpha (J_\alpha - J^\star_\alpha) \tag{10.31}$$

thus satisfies a simple evolution law under dilatation

$$K'^{(i)} = \lambda^{(i)}(l) K^{(i)}. \tag{10.32}$$

The dilatations satisfy the obvious composition rule

$$M(l_1 l_2) = M(l_1)M(l_2), \tag{10.33}$$

which goes over to eigenvalues. Therefore, one can define the exponents y_i as

$$\lambda^{(i)}(l) = l^{y_i}. \tag{10.34}$$

Although there is no a priori reason that guarantees that the $\lambda^{(i)}$ are real, in cases of interest here they are real. We must then distinguish three classes of interaction $K^{(i)}$:

- Interactions with negative y_i: they are *irrelevant*. Indeed, the iteration of the transformation drives them to zero. The large distance behaviour will not depend upon these variables. We can keep them, or set the corresponding $K^{(i)}$ to zero; it does not matter, they have no effect on the critical behaviour.
- Interactions with positive y_i: they are *relevant*. The corresponding $K^{(i)}$ increase during the iteration and, unless these relevant variables are set to zero, they drive the trajectories out of the critical region.
- Interactions with $y_i = 0$: they are called *marginal*. Their influence cannot be determined by the linearized transformations that we have considered.

The *critical manifold* (Fig. 10.6) in the p-dimensional space of interactions that we have considered is defined by setting to zero the k relevant (and marginal if any) variables. It has thus the dimension $(p-k)$. On this manifold the iteration, reduced now to the $(p-k)$ irrelevant variables, leads to the origin, i.e., the fixed point. For any point on this critical surface, the correlation length is infinite. Outside of the linearized domain that we have considered, the critical manifold is the basin of attraction of the fixed point.

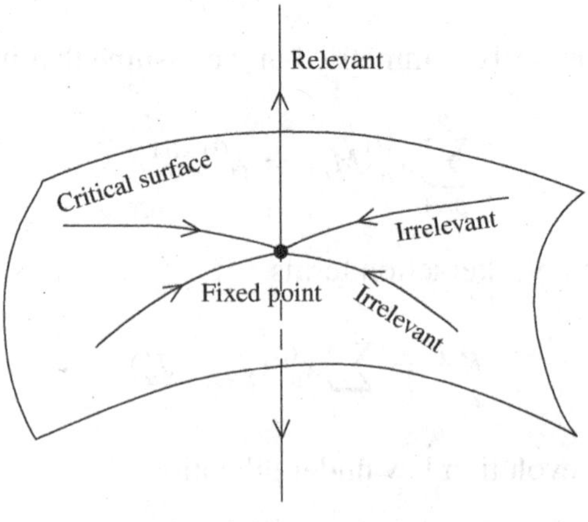

Figure 10.6 The critical surface, fixed points, relevant and irrelevant variables

However, in the directions defined by the eigenvectors with positive y_i, the fixed point is repulsive. The correlation length decreases when one iterates in such directions, one moves away from criticality. Therefore, unless the k relevant variables are fixed to zero, the behaviour of the system is not critical: the correlation length remains finite.

What can one say in a spin model, such as Ising? We know that it is sufficient to fix two parameters of the interaction to a specified value to find a critical point; indeed the two parameters are

$$J_1 = \beta h \qquad\qquad J_2 = \beta J. \tag{10.35}$$

The critical values are $J_1 = 0$, $J_2 = J/kT_c$. So even if the renormalization group transformation forces us to introduce p parameters instead of two, we know a priori that there can only be two relevant directions and a critical surface of dimension $(p-2)$. Indeed, if we had a third relevant variable among the p that we consider it would mean that a third physical quantity would have to be fixed to some specified value, to get a critical point: it would not be sufficient to set $h = 0$, $T = T_c$.

There are more complicated structures in nature with three relevant variables instead of two. For instance ternary mixtures may exhibit a tricritical point at some temperature and for two special composition ratios (three variables). This discussion may easily be carried over to such cases.

For an Ising-like model with Z_2 symmetry, the two relevant linear combinations are respectively even and odd under the group. The odd one is called magnetic (as in the coupling $\beta h\sigma$); the even one is called thermal (as in $\beta J\sigma_i\sigma_j$). Let us call the two corresponding relevant $K^{(i)}$, h and t which both vanish at T_c; therefore, $t \sim (T - T_c)$. The corresponding exponents are called y_h and y_T. We thus have the homogeneous scaling relation

$$\xi(hl^{y_h}, tl^{y_T}) = \frac{\xi(h,t)}{l}. \tag{10.36}$$

For the free energy per site, the scaling relation is inhomogeneous, as may be seen in the decimation example (10.18)

$$f(h,t) = \frac{f(hl^{y_h}, tl^{y_T})}{l^d} + g(h,t) \tag{10.37}$$

(since the number of sites is divided by l^d in the rescaling). The additional constant $g(h,t)$ is a non-universal non-singular function in the critical domain, since it involves only a local elimination of degrees of freedom. It does not contribute to the singular part of the free energy.

All the scaling laws and exponents defined in Chapter 7 follow from these relations. For instance, if we set $h = 0$ one finds (the reader should verify)

$$\nu = \frac{1}{y_T} \tag{10.38}$$

and after taking a derivative of (10.37) with respect to h

$$\frac{\gamma}{\nu} = d - 2y_h. \tag{10.39}$$

Summary

The renormalization group equations determine, whenever one can write the effect of a dilatation conveniently, the fixed points and the critical surface. The linearized flow near the fixed point allows one to compute the critical exponents. The only thing that we have not done yet is to construct explicitly these renormalization group transformations. To do it, beyond the hand waving of this chapter, we have to return to field theory.

11

Renormalization group for the φ^4 theory

Before moving on to the renormalized theory let us start in a more elementary way. We begin with the Boltzmann weight e^{-A}, with

$$A\{\varphi\} = \int d^d x \left\{ \frac{1}{2}(\vec{\nabla}\varphi)^2 + \frac{1}{2} r_0 \varphi^2 + \frac{g_0}{4!}\varphi^4 \right\} \tag{11.1}$$

and remember that there is an underlying short-distance cut-off, such as a lattice spacing. For simplicity, we begin with an $n = 1$ order parameter, and deal with $n > 1$ in Section 11.4.

In Chapter 8 we have computed the first correction to the mean-field free energy

$$\Gamma^{(1)}\{m\} = \int d^d r \left\{ \frac{1}{2}(\nabla m)^2 + \frac{1}{2} r_0(T) m^2 + \frac{g_0}{4!} m^4 \right\} + \frac{1}{2} \mathrm{Tr} \log K, \tag{11.2}$$

with

$$K(x, y) = \delta(x - y) \left(-\nabla^2 + r_0 + \frac{1}{2} g_0 m^2(x) \right). \tag{11.3}$$

This reduces, for a uniform field $h(x) = h$ and hence $m(x) = m$, to a free energy per unit volume

$$\gamma^{(1)}(m) = \frac{1}{2} r_0(T) m^2 + \frac{g_0}{4!} m^4 + \frac{1}{2} \int \frac{d^d q}{(2\pi)^d} \log \left(\frac{r_0 + \frac{1}{2} g_0 m^2 + \vec{q}^{\,2}}{r_0 + \vec{q}^{\,2}} \right), \tag{11.4}$$

where a constant has been subtracted so that $\gamma^{(1)}(m)$ vanishes with m. I have argued earlier that this expansion was useless near T_c. However, we shall see that it provides a method of computing the renormalization group trajectories under a change of scale.

Let us show that a change of scale, $a \to la$, or rather here $a \to \lambda^{-1} a$ ($\lambda < 1$) (and thus for the UV cut-off $\Lambda \to \lambda\Lambda$), can be accompanied by a modification of the parameters r_0 and g_0, which enter into A, while the physical content of the theory remains invariant.

11.1 Renormalization group … without renormalization

One can work with the theory as it is: the logic is simple, but the calculations are slightly more cumbersome than if we used the full machinery of renormalization theory.

Susceptibility

The magnetic susceptibility in zero field, above T_c, is

$$r = \chi^{-1} = \left.\frac{\partial^2 \gamma(m)}{\partial m^2}\right|_{m=0}. \tag{11.5}$$

Without fluctuations, r was just equal to r_0 and the critical temperature, defined as the value for which χ diverges, was defined by $r_0(T_c) = 0$.

With fluctuations,

$$r = r_0 + \frac{g_0}{2}\int \frac{\mathrm{d}^d q}{(2\pi)^d}\frac{1}{q^2 + r_0}, \tag{11.6}$$

and χ diverges for a different value of r_0, given by

$$0 = r_0^c + \frac{g_0}{2}\int \frac{\mathrm{d}^d q}{(2\pi)^d}\frac{1}{q^2 + r_0^c}. \tag{11.7}$$

In the logic of perturbation theory, we can invert these relations as

$$r_0 = r - \frac{g_0}{2}\int \frac{\mathrm{d}^d q}{(2\pi)^d}\frac{1}{q^2 + r}, \tag{11.8}$$

$$r_0^c = -\frac{g_0}{2}\int \frac{\mathrm{d}^d q}{(2\pi)^d}\frac{1}{q^2}. \tag{11.9}$$

Subtracting the two, we obtain

$$t = r_0 - r_0^c = r\left\{1 + \frac{g_0}{2}\int \frac{\mathrm{d}^d q}{(2\pi)^d}\frac{1}{q^2(q^2 + r)}\right\}. \tag{11.10}$$

We have seen earlier that in the critical limit $r \to 0$ the last integral blows up in the infrared (i.e., small q) for $d \leq 4$. The divergence is only logarithmic in four dimensions, more severe below four. So we begin with $d = 4$ and examine $d < 4$ in Section 11.5. There,

$$\int \frac{\mathrm{d}^4 q}{(2\pi)^4}\frac{1}{q^2(q^2 + r)} = \frac{2\pi^2}{(2\pi)^4}\int_0^{\Lambda} \frac{q^3 \mathrm{d}q}{q^2(q^2 + r)} \simeq \frac{2\pi^2}{(2\pi)^4}\log\frac{\Lambda}{\sqrt{r}}, \tag{11.11}$$

i.e.,

$$t = r\left\{1 + \frac{g_0}{2}\frac{2\pi^2}{(2\pi)^4}\log\frac{\Lambda}{\sqrt{r}}\right\}. \tag{11.12}$$

Coupling constant

Similarly, the fourth derivative of $\gamma(m)$ at $m = 0$ was g_0 in mean field theory, but this non-linear susceptibility becomes here

$$g = \left.\frac{\partial^4 \gamma(m)}{\partial m^4}\right|_{m=0} = g_0 - \frac{3g_0^2}{2} \int \frac{\mathrm{d}^4 q}{(2\pi)^4} \frac{1}{(q^2 + r_0)^2}. \tag{11.13}$$

This can also be perturbatively inverted:

$$\begin{aligned} g_0 &= g + \frac{3g^2}{2} \int \frac{\mathrm{d}^4 q}{(2\pi)^4} \frac{1}{(q^2 + r)^2} \\ &\simeq g + \frac{3g^2}{2} \frac{2\pi^2}{(2\pi)^4} \log \frac{\Lambda}{\sqrt{r}}, \end{aligned} \tag{11.14}$$

in which we have used $\Lambda^2 \gg r$. In the same logic, we replace (11.12) with

$$t = r \left\{ 1 + \frac{g}{2} \frac{2\pi^2}{(2\pi)^4} \log \frac{\Lambda}{\sqrt{r}} \right\}. \tag{11.15}$$

Renormalization group flow

We are now in a position to examine how a change of the parameters that entered the initial weight e^{-A}, when we rescale Λ, namely r_0 and g_0, may lead to the same physics. The physics is the free energy, not A; the derivatives of the free energy, such as r and g, are physically measurable quantities, whereas r_0 and g_0 are mere parameters to describe the microscopic interactions.

We are thus looking for a change of the microscopic parameters r_0, g_0 to $r_0(\lambda), g_0(\lambda)$, when the (inverse) length scale goes from Λ to $\lambda\Lambda$, while the physical parameters, such as r and g, remain invariant: several 'microscopic models' at different length scales thus provide the same physics. The explicit expressions (11.14, 11.15) make this very simple:

$$\begin{aligned} g_0(\lambda) &= g + \frac{3g^2}{2} \frac{2\pi^2}{(2\pi)^4} \log \frac{\lambda\Lambda}{\sqrt{r}} \\ &= g_0 + \frac{3g_0^2}{2} \frac{2\pi^2}{(2\pi)^4} \log \lambda, \end{aligned} \tag{11.16}$$

in which again we have used the freedom to replace g by g_0, the error being of higher order.

Similarly,

$$t(\lambda) = r \left\{ 1 + \frac{g}{2} \frac{2\pi^2}{(2\pi)^4} \log \frac{\lambda\Lambda}{\sqrt{r}} \right\} = t \left\{ 1 + \frac{g_0}{2} \frac{2\pi^2}{(2\pi)^4} \log \lambda \right\} \tag{11.17}$$

(using the same freedom).

It is customary to express the flow equations (11.16, 11.17) in a differential form:

$$\lambda \frac{\mathrm{d}g_0}{\mathrm{d}\lambda} = \beta(g_0) = \frac{2\pi^2}{(2\pi)^4}\frac{3}{2}g_0^2 + O(g_0^3). \tag{11.18}$$

It is thus simpler to redefine g_0 to get rid of the ubiquitous factor $\frac{2\pi^2}{(2\pi)^4}$ and write

$$\tilde{g}_0 = \frac{2\pi^2}{(2\pi)^4} g_0. \tag{11.19}$$

Then

$$\lambda \frac{\mathrm{d}\tilde{g}_0}{\mathrm{d}\lambda} = \beta(\tilde{g}_0) = \frac{3}{2}\tilde{g}_0^2 + O(\tilde{g}_0^3). \tag{11.20}$$

Similarly, from (11.17),

$$\lambda \frac{\mathrm{d}t}{\mathrm{d}\lambda} = t\kappa(\tilde{g}_0), \tag{11.21}$$

with

$$\kappa(g_0) = \frac{1}{2}\tilde{g}_0 + O(\tilde{g}_0^2). \tag{11.22}$$

(From now on we drop the tilde, $\tilde{g}_0 \to g_0$.)

We have seen that perturbation theory was meaningless if we tried to extract the critical behaviour from it. Nevertheless one can use it to study the renormalization group flow. We shall see that this flow provides the non-perturbative solution to the problem.

11.2 Study of the renormalization group flow in dimension four

We now want to study a limit in which the rescaling $l = 1/\lambda$ is large, i.e., $\lambda \to 0$. Let us look in general to the equation (11.20) for an arbitrary right-hand side $\beta(g_0)$:

$$\frac{\mathrm{d}g_0}{\beta(g_0)} = \mathrm{d}\log\lambda. \tag{11.23}$$

This shows that:

1. Whenever $\beta(g_0)$ is positive, g_0 decreases when λ decreases.
2. However, if $\beta(g_0)$ is negative, g_0 increases when λ decreases.
3. Consequently, when $\lambda \to 0$, $g_0(\lambda)$ tends to an infrared fixed point, i.e., a point $g_0^\star$, at which

$$\beta(g_0^\star) = 0, \quad \text{with} \quad \beta'(g_0^\star) > 0. \tag{11.24}$$

4. The origin is a fixed point, since $\beta(0) = 0$, but $\beta'(0)$ also vanishes: we are in the situation of a marginal variable, examined in the previous chapter: a linearization around the

fixed point is not sufficient. However, for small g_0, the equation (11.20) is elementary and shows that the origin is indeed an infrared attractor, since

$$g_0(\lambda) \simeq \frac{2}{3} \frac{1}{\log 1/\lambda}. \tag{11.25}$$

Let us also write (11.17) in differential form:

$$\lambda \frac{\mathrm{d}t}{\mathrm{d}\lambda} = t(\lambda)\kappa[g_0(\lambda)] \tag{11.26}$$

with

$$\kappa(g_0) = \frac{1}{2} g_0 + O(g_0^2), \tag{11.27}$$

which gives

$$t(\lambda) = t\mathrm{e}^{\int_1^\lambda \frac{\mathrm{d}\mu}{\mu}\kappa[g_0(\mu)]}. \tag{11.28}$$

If there is an infrared fixed point at a finite $g_0^\star$ then, when $\lambda \to 0$,

$$\int_1^\lambda \frac{\mathrm{d}\mu}{\mu}\kappa[g_0(\mu)] \sim \kappa[g_0^\star] \int_1^\lambda \frac{\mathrm{d}\mu}{\mu}, \tag{11.29}$$

and, in the limit of small λ, one has

$$t(\lambda) \sim t\lambda^{\kappa(g_0^\star)}. \tag{11.30}$$

However, for the case under consideration here of a fixed point at the origin, it is useful to write

$$\frac{\mathrm{d}\mu}{\mu} = \frac{\mathrm{d}g_0}{\beta(g_0)} \tag{11.31}$$

and thus we change variable from μ to $g_0(\mu)$,

$$\int_1^\lambda \frac{\mathrm{d}\mu}{\mu}\kappa[g_0(\mu)] = \int_{g_0}^{g_0(\lambda)} \frac{\kappa(g)}{\beta(g)}\mathrm{d}g. \tag{11.32}$$

Since here $\frac{\kappa}{\beta} = \frac{1}{3g}(1 + O(g))$,

$$\int_{g_0}^{g_0(\lambda)} \frac{\kappa(g)}{\beta(g)}\mathrm{d}g \sim \frac{1}{3} \log \frac{g_0(\lambda)}{g_0} \tag{11.33}$$

and finally, for small λ

$$t(\lambda) \sim t(\log 1/\lambda)^{-1/3}. \tag{11.34}$$

11.3 Critical behaviour of the susceptibility in dimension four

Let us recall the problem that we encountered when we tried perturbation theory. At the lowest order (mean field), we found

$$\chi^{-1} \sim t. \tag{11.35}$$

At the next order we had

$$\chi^{-1} \sim t\left(1 - \frac{1}{2}g_0 \log \Lambda \Big/ \sqrt{t}\right). \tag{11.36}$$

The alleged 'correction' was divergent, and the next term would have given a $\log^2$, and so on.

In the previous section, we have used perturbation theory not to compute χ, but to determine the renormalization group flow equations for small coupling. We have found that the physical quantities, such as the susceptibility, were invariant when we changed Λ to $\lambda\Lambda$ (i.e., $a \to a/\lambda$), provided we replaced g_0 and t by $g_0(\lambda)$ and $t(\lambda)$:

$$\chi^{-1}(t, g_0, \Lambda) = \chi^{-1}(t(\lambda), g_0(\lambda), \lambda\Lambda). \tag{11.37}$$

Remark

We have not taken into account the renormalization of the order parameter itself, since it begins only at next order (two-loop order), and all our calculations are limited here to one loop. If we went further, but as we shall see it is not necessary here, we should include a rescaling of the order parameter.

We now add to this an *ordinary dimensional analysis*, in which we take as unit an inverse length such as Λ. In these units $[\chi^{-1}] = 2,\ [t] = 2,\ [g_0] = 0$. Consequently,

$$\chi^{-1}(t(\lambda), g_0(\lambda), \lambda\Lambda) = \lambda^2 \chi^{-1}\left(\frac{t(\lambda)}{\lambda^2}, g_0(\lambda), \Lambda\right). \tag{11.38}$$

This equation is completely trivial, contrary to (11.37), which contains the essence of the renormalization group flow.

Combining the two equations, we find

$$\chi^{-1}(t, g_0, \Lambda) = \lambda^2 \chi^{-1}\left(\frac{t(\lambda)}{\lambda^2}, g_0(\lambda), \Lambda\right). \tag{11.39}$$

The problem we had with perturbation theory came from the domain in which $t \to 0$ in the scale Λ^2, in which it became meaningless. Let us use the arbitrariness of the dilatation parameter λ to go in the r.h.s. of (11.39) to a domain in which $\frac{t(\lambda)}{\lambda^2}$ remains finite in the same scale, Λ^2. For instance, we chose the dilatation $\lambda = \lambda^*$ so that

$$\frac{t(\lambda)}{(\lambda^* \Lambda)^2} = 1, \tag{11.40}$$

which is the equation that we are going to solve for λ^*, and show that indeed λ^* fixed by this condition goes to zero in the critical domain.

Once this is done, (11.39) will take the form

$$\chi^{-1}(t, g_0, \Lambda) = (\lambda^*)^2 \chi^{-1}(\Lambda^2, g_0(\lambda^*), \Lambda). \tag{11.41}$$

Assume again that λ^* is going to zero; then $\chi^{-1}(\Lambda^2, g_0(\lambda^*), \Lambda)$ is very simple: its ratio t/Λ^2, which led to the breakdown of perturbation theory is equal to one; the coupling constant is small. *We have thus achieved a dual goal: while moving away under dilatation from the critical region in which perturbation theory failed, we have generated at the same time a small coupling constant at the new scale.* Perturbation theory is now perfectly legitimate and, returning to (11.36), we see that

$$\chi^{-1}(\Lambda^2, g_0(\lambda^*), \Lambda) = \Lambda^2 \left(1 + O\left(\frac{1}{\log 1/\lambda^*}\right)\right). \tag{11.42}$$

We have assumed, but not yet shown, that $\lambda^* \to 0$. Let us assume it again and show that it is consistent: we have found in (11.34) that $t(\lambda) \sim t(\log 1/\lambda)^{-1/3}$. Then (11.40) yields

$$(\lambda^*)^2 (\log 1/\lambda^*)^{1/3} = \frac{t}{\Lambda^2}, \tag{11.43}$$

whose solution is asymptotically (i.e., for t much smaller than Λ^2)

$$(\lambda^*)^2 = \frac{t}{\Lambda^2} \frac{2}{(\log t/\Lambda^2)^{1/3}}. \tag{11.44}$$

Returning to (11.41) and (11.42) we end up with

$$\chi^{-1}(t, g_0, \Lambda) \simeq_{t \to 0} C t (\log t/\Lambda^2)^{-1/3}, \tag{11.45}$$

a result that would have been difficult to guess with perturbation theory!

The final result (11.45) is universal; it does not depend on how we put the cut-off, it does not depend on g_0, which was a priori marginal, but finally was driven as an inverse logarithm to zero.

We have used perturbation theory to compute the first loop contribution to the renormalization group flow. What would have happened if we had computed higher orders? Nothing, if the coupling constant remains in the domain of attraction of the origin $g_0 = 0$. Are there other fixed points? This is a non-perturbative question. Numerical simulations (Monte-Carlo method) for four-dimensional models have never found a fixed point other than the origin.

Of course, there is no way to do an experiment in four space dimensions. However, Larkin and Khmel'nitskii showed in 1969 that uniaxial dipolar ferromagnets with strong dipole–dipole interactions (which are long range) should exhibit in three dimensions the same behaviour as this four-dimensional model with short-range interactions.[1] This has given rise to very precise experimental tests;[2] G. Ahlers and co-workers have worked with the dipolar Ising ferromagnet $LiTbF_4$ (i.e., with one direction of easy magnetization); the Curie temperature $T_c \simeq 2.9$ K. They have measured the specific heat near T_c. Landau theory predicts a discontinuity at T_c, whereas the renormalization group predicts, for the two sides of T_c:

$$C_\pm = A_\pm (\log 1/|t|)^p, \quad \text{with} \quad p = \frac{1}{3} \quad \text{and} \quad \frac{A_+}{A_-} = \frac{1}{4}, \tag{11.46}$$

with $t = (T - T_c)/T_c$. The experiment does show a weak divergence of the specific heat at T_c, and is fitted with a power of the logarithm $p = 0.34 \pm 0.03$; the amplitude ratio is found to be $A_+/A_- = 0.24 \pm 0.01$. The agreement with the theory is impressive and one should note that here, since the fixed point is at the origin, the theory does not involve any approximation procedure, such as epsilon expansion or anything: it is a pure test of the renormalization group ideas. This is, in fact, very similar in spirit to the tests of quantum chromodynamics, based on logarithmic deviations to scaling at high energy.

11.4 Multi-component order parameters

The method that was exposed in Chapter 9, which led to the loop expansion of the free energy (9.15), is known as the background field method. It is particularly well suited to problems with more complex symmetries than Z_2.

Let us briefly see what would happen if we had an n-component field with $O(n)$ symmetry. If we had used Feynman diagrams, instead of the method exposed here, we would have to follow with care the flow of internal indices in the diagram, given that the interaction terms are $\sum_{\alpha\beta} \varphi_\alpha\varphi_\alpha\varphi_\beta\varphi_\beta$ and the propagator is proportional to $\delta_{\alpha\beta}$.

In the method that we have chosen, we have

$$A\{\varphi\} - \int \mathrm{d}^d r\, \mathbf{h}(\vec{r}) \cdot \varphi(\vec{r}) = A\{\hat{\varphi}\} - \int \mathrm{d}^d r\, \mathbf{h}(\vec{r}) \cdot \hat{\varphi}(\vec{r})$$
$$+ \frac{1}{2} \int \mathrm{d}^d r \mathrm{d}^d s\, \psi_\alpha(\vec{r}) \left. \frac{\delta^2 A}{\delta\varphi_\alpha(\vec{r})\delta\varphi_\beta(\vec{s})} \right|_{\hat{\varphi}} \psi_\beta(\vec{s}) + O(\psi^3). \tag{11.47}$$

[1] A. I. Larkin and D. E. Khmel'nitskii, *Zh. Eksp. Teor. Fiz.*, **56** (1969) 2087 [*Sov. Phys. JETP*, **29** (1969) 1123]. See also E. Brézin and J. Zinn-Justin, *Phys. Rev., B*, **13** (1976) 251.

[2] G. Ahlers, A. Kornblit and H. Guggenheim, *Phys. Rev. Lett.*, **34** (1975) 1227.

Then, since

$$\frac{\delta A}{\delta \varphi_\alpha(\vec{r})} = \left(-\nabla^2 + r_0 + \frac{g_0}{6}\vec{\varphi}^2\right)\varphi_\alpha, \tag{11.48}$$

the second derivative of A is given by the kernel

$$\begin{aligned} K_{\alpha,\beta}(\vec{r},\vec{s}) &= \left.\frac{\delta^2 A}{\delta\varphi_\alpha(\vec{r})\delta\varphi_\beta(\vec{s})}\right|_{\hat{\varphi}} \\ &= \delta(\vec{r}-\vec{s})\left(-\nabla^2 + r_0 + \frac{g_0}{6}\vec{\varphi}^2\right)\delta_{\alpha\beta} + \left.\frac{g_0}{3}\varphi_\alpha\varphi_\beta\right|_{\varphi=m}. \end{aligned} \tag{11.49}$$

As earlier, the diagonalization is done by splitting the longitudinal and transverse directions to the vector $m_\alpha = m e_\alpha$ in which e is the unit vector along the magnetization. This is done as follows:

$$\begin{aligned} K_{\alpha,\beta}(\vec{r},\vec{s}) &= \delta(\vec{r}-\vec{s})\left(-\nabla^2 + r_0 + \frac{g_0}{6}m^2\right)(\delta_{\alpha\beta} - e_\alpha e_\beta) \\ &\quad + \delta(\vec{r}-\vec{s})\left(-\nabla^2 + r_0 + \frac{g_0}{2}m^2\right)e_\alpha e_\beta. \end{aligned} \tag{11.50}$$

As far as the internal space is concerned, we have one longitudinal kernel,

$$K_{\mathrm{L}}(\vec{r},\vec{s}) = \delta(\vec{r}-\vec{s})\left(-\nabla^2 + r_0 + \frac{g_0}{2}m^2\right), \tag{11.51}$$

and an $(n-1)$-fold degenerate transverse kernel,

$$K_{\mathrm{T}}(\vec{r},\vec{s}) = \delta(\vec{r}-\vec{s})\left(-\nabla^2 + r_0 + \frac{g_0}{6}m^2\right). \tag{11.52}$$

Therefore, the formula (11.4), for a uniform magnetization m, becomes

$$\begin{aligned} \operatorname{Tr}\log K &= \frac{V}{(2\pi)^d}\int \mathrm{d}^d q \log\left(r_0 + \frac{g_0}{2}\, m^2 + \vec{q}^2\right) \\ &\quad + (n-1)\frac{V}{(2\pi)^d}\int \mathrm{d}^d q \log\left(r_0 + \frac{g_0}{6}\, m^2 + \vec{q}^2\right). \end{aligned} \tag{11.53}$$

The equation (11.13) is thus replaced by

$$g = g_0 - \frac{3g_0^2}{2}\int \frac{\mathrm{d}^d q}{(2\pi)^d}\frac{1}{(q^2+r_0)^2} - (n-1)\frac{g_0^2}{3}\int \frac{\mathrm{d}^d q}{(2\pi)^d}\frac{1}{(q^2+r_0)^2}. \tag{11.54}$$

The four-dimensional flow equation (11.45) is replaced by

$$g_0(\lambda) = g_0 + g_0^2\left(\frac{3}{2} + \frac{(n-1)}{3}\right)\frac{2\pi^2}{(2\pi)^4}\log\lambda \tag{11.55}$$

and thus the beta function becomes

$$\beta(g_0) = \frac{2\pi^2}{(2\pi)^4}\frac{n+8}{6}g_0^2 + O(g_0^3). \tag{11.56}$$

Similarly, for the flow of temperature, one finds

$$t(\lambda) = t\left\{1 + g_0 \frac{n+2}{6} \frac{2\pi^2}{(2\pi)^4} \log \lambda\right\}. \tag{11.57}$$

We can thus repeat for the n-component order parameter every step that we had done before for $n = 1$.

Remark

Up to now, we have used the free energy functional (11.2) only for a uniform order parameter $m(x) = m$. However, the full free energy functional $\Gamma\{m\}$ is in fact a generating function for all the irreducible correlation functions

$$\Gamma^{(k)}(x_1, \ldots, x_k) = \frac{\delta}{\delta m(x_1)} \cdots \frac{\delta}{\delta m(x_k)} \Gamma\{m\}. \tag{11.58}$$

The name irreducible comes from the fact that in perturbation theory the functions are represented by diagrams that cannot be split into two pieces by cutting along one single line. In particular, for $k = 2$, $\Gamma^{(2)}(x, y)$ is the inverse (as operator) of the two-point correlation function.

11.5 Epsilon expansion

In dimensions other than four, the coupling constant is not dimensionless. Since $A\{\varphi\}$ is dimensionless (the Boltzmann weight is e^{-A}),

$$A\{\varphi\} = \int \mathrm{d}^d r \left\{\frac{1}{2}(\nabla\varphi)^2 + \frac{1}{2} r(T)\varphi^2 + \frac{g_0}{4!}\varphi^4\right\}. \tag{11.59}$$

in a system of units in which a length has dimension -1, one sees that the dimension of $\varphi = \frac{1}{2}(d-2)$ and thus $[g_0] = 4 - d$. Therefore, we will extract this dimension to work with a dimensionless coupling constant, and for doing that we have to examine the main length scales of the problem: we have a the lattice spacing and ξ the correlation length. But g_0 is microscopic, it is just related to the spin distribution at one site, therefore, it cannot depend on a length such as ξ, which results from the spin correlations. Therefore, its dimension is given by a or, here, its inverse Λ: we thus change

$$g_0 \to \Lambda^{\epsilon} g_0, \tag{11.60}$$

in which $\epsilon = 4 - d$ and the new g_0 is dimensionless.

For reasons that will be transparent further, we shall assume that we expand both in powers of g_0 and ϵ, the two being of the same order of magnitude. Therefore, wherever we neglected g_0^2 we can also neglect $g_0\epsilon$. Returning to our previous equation (11.14), nothing is changed except the new scale in the l.h.s.:

$$\Lambda^\epsilon g_0 = g + \frac{3g^2}{2}\int \frac{\mathrm{d}^d q}{(2\pi)^d}\frac{1}{(q^2+r)^2}$$
$$\simeq g + \frac{3g^2}{2}K_d \log\frac{\Lambda}{\sqrt{r}}, \tag{11.61}$$

in which K_d is the 'area' of the unit sphere in dimension d, divided by $(2\pi)^d$,

$$K_d = \frac{2\pi^{d/2}}{\Gamma(d/2)(2\pi)^d}. \tag{11.62}$$

This factor is now absorbed into a redefinition of g_0 and g:

$$g_0 K_d \to g_0 \qquad\qquad g K_d \to g. \tag{11.63}$$

If we change the scale $\Lambda \to \lambda\Lambda$ as before, i.e., by fixing the physical parameters of the free energy g, r, we have one new term in the l.h.s., namely

$$\Lambda^\epsilon = 1 + \epsilon\log\Lambda + O(\epsilon^2), \tag{11.64}$$

then (11.61) becomes

$$\lambda^\epsilon\Lambda^\epsilon g_0(\lambda) = g + \frac{3g^2}{2}\log\frac{\lambda\Lambda}{\sqrt{r}}$$
$$\simeq g_0\Lambda^\epsilon + \frac{3g_0^2}{2}\log\lambda. \tag{11.65}$$

Neglecting terms of order $g_0\epsilon^2$, this gives

$$g_0(\lambda) = g_0(1-\epsilon\log\lambda) + \frac{3g_0^2}{2}\log\lambda, \tag{11.66}$$

and thus

$$\beta(g_0) = \lambda\frac{\mathrm{d}g_0}{\mathrm{d}\lambda} = -\epsilon g_0 + \frac{3g_0^2}{2}\log\lambda + O(g_0^3, g_0^2\epsilon, g_0\epsilon^2) \tag{11.67}$$

and for n components one would find instead (a simple recommended exercise)

$$\beta(g_0) = \lambda\frac{\mathrm{d}g_0}{\mathrm{d}\lambda} = -\epsilon g_0 + \frac{(n+8)}{6}g_0^2 + O(g_0^3, g_0^2\epsilon, g_0\epsilon^2). \tag{11.68}$$

The infrared fixed point (at which the derivative of beta is positive) has moved away from the origin and is now

$$g_0^\star = \frac{6}{n+8}\epsilon + O(\epsilon^2), \tag{11.69}$$

justifying the fact that we treated g_0 and ϵ as small parameters of the same order.

This is the basis of the Wilson–Fisher epsilon expansion.[3] Let us examine the consequences of this new flow. For instance, the correlation length ξ, a physically calculable quantity in terms of g and r, which have been kept fixed, satisfies

$$\xi(g_0, t, \Lambda) = \xi(g_0(\lambda), t(\lambda), \lambda\Lambda) = \frac{1}{\lambda}\xi(g_0(\lambda), t(\lambda)/\lambda^2, \Lambda), \tag{11.70}$$

where we have used the renormalization group, followed by elementary dimensional analysis.

As done earlier in four dimensions, we use a dilatation $\lambda \to 0$, so that the dimensionless ratio $t(\lambda)/(\Lambda\lambda)^2$ is not small and we fix it to a finite number, for instance, we choose λ so that

$$\frac{t(\lambda^*)}{(\lambda^*\Lambda)^2} = 1. \tag{11.71}$$

This, if λ^* is small (and we shall verify it soon) drives $g_0(\lambda^*)$ to the infrared fixed point (11.69) and we have seen in (11.30) that

$$t(\lambda^*) \sim (\lambda^*)^{\kappa(g_0^\star)} t. \tag{11.72}$$

This gives

$$\lambda^* = \left(\frac{t}{\Lambda^2}\right)^{\frac{1}{2-\kappa(g_0^\star)}}, \tag{11.73}$$

which is indeed small in the critical region. We have thus

$$\xi(g_0, t, \Lambda) = \frac{1}{\lambda^*}\xi(g_0^\star, \Lambda^2, \Lambda) \sim \left(\frac{t}{\Lambda^2}\right)^{-\frac{1}{2-\kappa(g_0^\star)}}. \tag{11.74}$$

This shows that the exponent ν is given by

$$\nu = \frac{1}{2-\kappa(g_0^\star)}, \tag{11.75}$$

and taking the function κ that we calculated earlier,

$$\kappa(g_0) = \frac{1}{2} g_0 + O(g_0^2), \tag{11.76}$$

for $n = 1$ and for arbitrary n,

$$\kappa(g_0) = \frac{n+2}{6} g_0 + O(g_0^2), \tag{11.77}$$

we have

$$\kappa(g_0^\star) = \frac{n+2}{n+8}\epsilon + O(\epsilon^2), \tag{11.78}$$

[3] K. G. Wilson and M. E. Fisher, Critical exponents in 3.99 dimensions, *Phys. Rev. Lett.*, **28** (1972) 240.

and thus

$$\nu = \frac{1}{2} + \frac{n+2}{4(n+8)}\epsilon + O(\epsilon^2). \tag{11.79}$$

For $n = 1$ and $\epsilon = 1$ (three dimensions) this gives $\nu \simeq 0.58$, a better result than the mean field result 0.5, but still not the experimental (numerical) value 0.63.

These calculations can be pushed to higher orders in ϵ, but it is much easier at higher orders to use the renormalized massless theory. The calculations have been made up to order ϵ^5, a hard job. The results taken merely by setting $\epsilon = 1$ are useless, in the sense that the inclusion of the order ϵ^2 gives a value closer to 0.63, but the next terms are larger and larger, with alternating signs, revealing an asymptotic but divergent series, and one cannot use the expansion without fancier numerical methods. The simplest is a Padé approximant, but there are very elaborate methods,[4] which give the exponents with an accuracy of 10^{-3}. Such methods are able to transform the divergent series into a seemingly convergent algorithm; thus if one needed a more accurate value from the theory, assuming that experiments determine the exponents with an accuracy better than 10^{-3}, then we would have to calculate one more term in the ϵ expansion, painful but not an a priori obstacle. The situation is similar to that of accurate calculations in quantum electrodynamics, such as the magnetic moment of the electrons.

11.6 An exercise on the renormalization group: the cubic fixed point

To illustrate the structure of the renormalization group approach in other situations, let us examine the influence of a *cubic anisotropy*, i.e., of an additional coupling for an n-component order parameter φ_a, of the form $\sum_1^n \varphi_a^4$. Such a term breaks the $O(n)$ symmetry down to the permutation group H plus the symmetry $\varphi_a \to -\varphi_a$, i.e., $Z_2^n \times S_n$. The Boltzmann weight in this case is

$$A\{\varphi\} = \int \mathrm{d}^d x \left\{ \frac{1}{2}(\nabla\vec{\varphi})^2 + \frac{1}{2} r_0 \vec{\varphi}^2 + \frac{u_0}{4!}(\vec{\varphi}^2)^2 + \frac{v_0}{4!}\sum_1^n \varphi_a^4 \right\}. \tag{11.80}$$

At mean field level, the magnetization is related to the (uniform) applied field by

$$h_a = r_0 m_a + \frac{u_0}{6}\vec{m}^2 m_a + \frac{v_0}{6} m_a^3 \tag{11.81}$$

and, if the field vanishes, either $m_a = 0$ or

$$r_0 + u_0 \vec{m}^2 + \frac{v_0}{6} m_a^2 = 0. \tag{11.82}$$

[4] J. C. Le Guillou and J. Zinn-Justin, *J. Physique Lett.*, **46** (1985) L137.

In the region of spontaneous symmetry breaking ($r < 0$), note that all the non-vanishing components of $\vec{m}$ are equal or opposite, according to the sign of the components of $\vec{h}$, which we choose to let go to zero by positive values. It is left to the reader to verify that the free energy is minimum if all the components of the magnetization are non-zero; therefore the mean-field spontaneous magnetization is

$$m_a = m\vec{e} \tag{11.83}$$

in which $\vec{e}$ is the unit vector

$$\vec{e} = \frac{1}{\sqrt{n}}(1, \ldots, 1) \tag{11.84}$$

and

$$m^2 = -\frac{6r_0}{u_0 + v_0/n}. \tag{11.85}$$

The calculation of the one-loop fluctuations follows the same lines:

$$\begin{aligned} K_{ab}(\vec{r}, \vec{s}) &= \frac{\delta^2 A}{\delta\varphi_a(\vec{r})\delta\varphi_b(\vec{s})} \\ &= \delta(\vec{r} - \vec{s})\left(-\nabla^2 + r_0 + \frac{u_0}{6}\vec{\varphi}^2\right)\delta_{ab} + \frac{u_0}{3}\varphi_a\varphi_b + \frac{v_0}{2}\delta_{ab}\varphi_a^2 \end{aligned} \tag{11.86}$$

and for a constant magnetization one can collect the terms of degree four in φ in $\operatorname{Tr}\log K$. The relation (11.13) is now replaced by a pair of equations

$$u = u_0 - \left(\frac{n+8}{6}u_0^2 + u_0 v_0\right)\int \frac{d^4q}{(2\pi)^4}\frac{1}{(q^2 + r_0)^2} \tag{11.87}$$

and

$$v = v_0 - \left(\frac{3}{2}v_0^2 + 2u_0 v_0\right)\int \frac{d^4q}{(2\pi)^4}\frac{1}{(q^2 + r_0)^2}. \tag{11.88}$$

We then obtain two beta functions in dimension $4 - \epsilon$, replacing (11.68), namely

$$\beta_u(u_0, v_0) = \lambda\frac{du_0}{d\lambda} = -\epsilon u_0 + \frac{(n+8)}{6}u_0^2 + u_0 v_0, \tag{11.89}$$

$$\beta_v(u_0, v_0) = \lambda\frac{dv_0}{d\lambda} = -\epsilon v_0 + \frac{3}{2}v_0^2 + 2u_0 v_0, \tag{11.90}$$

in which we have neglected all cubic terms. There are three types of non-vanishing fixed points, at which both beta functions vanish:

(a) $v_0^\star = 0,\ u_0^\star = 6\epsilon/(n+8) + O(\epsilon^2)$;
clearly one is back to the $O(n)$ vector model without cubic anisotropy,

(b) $u_0^\star = 0,\ v_0^\star = 2\epsilon/3 + O(\epsilon^2)$;
in which the n components of φ_a are decoupled and lead to the $n = 1$ critical behaviour.

(c) $u_0^\star = 2\epsilon/n$, $v_0^\star = \frac{2}{3}(1-4/n)\epsilon$;
the new 'cubic' fixed point.

Which of these three fixed points is physically relevant? In the long-distance limit one flows to an infrared attractor, which corresponded to a positive derivative of the beta function at the fixed point, for the case of a single coupling constant. In the case of several coupling constants g_α, the flow near a fixed point $g_\alpha^\star$ is given by

$$\lambda\frac{\mathrm{d}}{\mathrm{d}\lambda}(g_\alpha - g_\alpha^\star) \simeq M_{\alpha\beta}(g_\alpha - g_\alpha^\star),$$

in which

$$M_{ab} = \left.\frac{\partial\beta_a}{\partial g_b}\right|_{g^\star}$$

and the fixed point is an infrared attractor if the eigenvalues of the matrix M are non-negative (or have a non-negative real part if they are complex). For this problem, the matrix M is given by

$$M = \begin{pmatrix} -\epsilon + \frac{n+8}{3}u_0 + v_0 & u_0 \\ 2v_0 & -\epsilon + 2u_0 + 3v_0 \end{pmatrix}. \tag{11.91}$$

At the $O(n)$-symmetric fixed point the eigenvalues are $(\epsilon, \frac{4-n}{n+8}\epsilon)$; at the decoupled fixed point they are $(-\epsilon/3, \epsilon)$; at the cubic fixed point they are $(\epsilon, \epsilon(n-4)/3n)$.

Finally: as long as $n < 4 + O(\epsilon)$, the stable fixed point is the $O(n)$-symmetric fixed point. However, if $n > 4+O(\epsilon)$, the cubic fixed point is stable and the critical behaviour is governed by a new set of critical exponents.[5]

Conclusion

The renormalization group has provided a complete theory of critical phenomena; it has explained universality and the scaling laws, and it has led to accurate calculations of the universal physical quantities.

[5] In fact any additional coupling breaking the $O(n)$-symmetry is irrelevant for $n < 4 + O(\epsilon)$, but it destabilizes the $O(n)$-symmetric fixed point for $n > 4$: E. Brézin, J.C. Le Guillou and J. Zinn-Justin, *Phys. Rev. B*, **10** (1974) 892.

12

Renormalized theory

The approach of Chapter 11, in which we have kept a large momentum or short-distance cut-off, is sufficient for the needs of computing the various universal indices and functions that characterize criticality. All calculations were made in the limit in which the ratio ξ/a of the correlation length ξ to the lattice spacing a is large. In the previous chapters, this was done by keeping a fixed and increasing ξ.

The renormalized theory relies on the opposite strategy: it deals directly with the scaling theory in which one takes the limit of a vanishingly small a. I have included here a chapter on the renormalized theory for several reasons:

- There is no ultraviolet (or lattice) cut-off any more and, getting rid of one dimensionful parameter considerably simplifies the integrations, especially at higher orders.
- All the critical exponents and scaling functions can be computed at T_c, i.e., in a massless theory, even when one considers temperature-dependent laws, such as those concerning the correlation length, the magnetic susceptibility or the spontaneous magnetization. Integrals in a massless theory without cut-off are by far simpler, and they are often given by mere dimensional analysis.
- However, the main reason is to understand the meaning of renormalizability from the viewpoint both of critical phenomena and of high-energy physics. This will be done in the following section.

12.1 The meaning of renormalizability

Renormalization theory was discovered for the needs of particle physics. Let us briefly recall the main points for a quantum field theory, such as quantum electrodynamics (QED). The early realization of the occurrence of divergences due to quantum fluctuations led to the necessity of 'regularizing' the theory, i.e., to the introduction of a cut-off, $\Lambda = a^{-1}$. Then it was shown that, provided one traded the bare parameters m_0 and e_0, the mass and charge of the electron, with the observed physical ones, as well as a change of the normalizations of the matter field and electromagnetic field, then the limit of all the physically measurable quantities existed if one let a go to zero: the theory is thus 'renormalizable'. This is true order by order in perturbation theory. However, this applied provided one dealt with one special well-defined Lagrangian, the one in which the 'minimal' coupling between the matter field and the electromagnetic field is induced by the replacement of the operator momentum of the electron p_μ by $p_\mu - eA_\mu$. Terms of higher degree in the fields, even if gauge invariant, such as a coupling between the spin-current to the electromagnetic field, would destroy the existence of this limit, namely the renormalizability.

If we remain with scalar fields, as was the case earlier in this book, the same is true of the φ^4 theory: shifting the bare Landau parameters r_0 and g_0 to the physically measurable ones r and g, and rescaling the field

$$\varphi \to \varphi = \sqrt{Z}\varphi_{\rm R} \tag{12.1}$$

allows one (up to four dimensions) to go to the limit in which ξ/a is infinite by letting $a \to 0$. However, this procedure is a priori not consistent with the point of view that we have followed up to now: we have started with a Landau functional of φ, which had no reason to be a polynomial of degree four. If we added a $g_6\varphi^6$ term, for instance, would that destroy renormalizability? Traditionally, the answer is indeed that a φ^6 (as well as any other higher degree term) completely spoils renormalizability in dimension four: indeed, simple power-counting arguments show that the degree of divergence increases with the order in the expansion in powers of the fluctuations. The 'elimination' of divergences thus requires the introduction of new parameters at each order and there is no theory any more. This seems to contradict completely the viewpoint that was developed earlier, in which terms of higher degree did not matter, and let us wonder why the scaling limit exists in spite of these higher terms.

There are two ways of understanding why those higher degree non-renormalizable terms do not matter:

- One follows the renormalization group flow, including the terms of higher degree, and one shows that they are irrelevant: i.e., whatever values the parameters, such as g_6 here, have initially, they flow to the fixed point when a becomes vanishingly small, namely $r = 0$ (or $T = T_c$), $g = g^\star$ (zero in dimension four, proportional to ϵ in dimension $d = 4 - \epsilon$) and $g_6^\star = 0$. Non-renormalizable terms have no effect when a goes to zero. This route was first introduced in the revolutionary article that followed K. Wilson's lectures in 1971–72[1] and developed as a full theory of renormalization by Polchinski.[2]

[1] K. G. Wilson and J. Kogut, The renormalization group, *Phys. Rep.*, **12c** (1974) 75.
[2] J. Polchinski, *Nucl. Phys. B*, **231** (1984) 269.

- Another way of looking at this irrelevance is as follows. Let us return to dimensional analysis; in the system of units in which [momentum = 1], [length = −1], we know that $[\varphi] = (d-2)/2 = 1$ in four dimensions, and thus, for the coefficient of φ^6, one finds $[g_6] = 3(d-2) - d$, namely dimension two when $d = 4$. In the unrenormalized theory, the parameter g_6 is of microscopic origin; it results from some coarse-grained Hamiltonian. Therefore, its dimension is related to the lattice spacing and not to the correlation length, which results from the full theory. In other words, the parameter g_6 should be taken as proportional to $\Lambda^{-2} = a^2$ multiplied by some dimensionless coupling constant, which we rename again g_6. We are thus really looking at the large Λ limit of a φ^4 cut-off theory with an additional interaction, $\frac{1}{\Lambda^2} g_6 \varphi^6$. It is then not difficult to examine the effect of this additional term as a perturbation: when Λ is large, its effect is simply to modify the value of the previous parameters r_0 and g_0 (as well as the normalization of the field). But since these parameters have already been included, and since the results that we have found are independent of the specific value of g_0 it does not lead to anything new. A full discussion on this point would be beyond the scope of this book, but to explain the point simply, let us return to the one-loop free energy that we considered in the previous chapters. We thus start now with

$$A\{\varphi\} = \int \mathrm{d}^d x \left\{ \frac{1}{2}(\vec{\nabla}\varphi)^2 + \frac{1}{2} r_0 \varphi^2 + \frac{g_0}{4!}\varphi^4 + \frac{1}{\Lambda^2}\frac{g_6}{6!}\varphi^6 \right\}. \tag{12.2}$$

The first correction to the mean-field free energy, for a constant external field, for instance, becomes

$$\gamma^{(1)}(m) = \frac{1}{2} r_0(T) m^2 + \frac{g_0}{4!} m^4 + \frac{1}{\Lambda^2}\frac{g_6}{6!} m^6 + \frac{1}{2}\int \frac{\mathrm{d}^4 q}{(2\pi)^4} \log\left(\frac{r_0 + \vec{q}^{\,2} + \frac{1}{2} g_0 m^2 + \frac{g_6}{24\Lambda^2} m^4}{r_0 + \vec{q}^{\,2}} \right). \tag{12.3}$$

Expanding this free energy in powers of m, we obtain new terms. Let us examine successively the coefficient of m^6 in the free energy, and the additional contribution to the coefficient of m^4 (see (11.13)).

The coefficient of m^6 in the free energy

We expand $\gamma^{(1)}$ up to order six; let us call the coefficient $G_6/6!$. One obtains

$$G_6 = \frac{g_6}{\Lambda^2} - \frac{15}{4}\frac{g_0 g_6}{\Lambda^2}\int \frac{\mathrm{d}^4 q}{(2\pi)^4}\frac{1}{(r_0 + q^2)^2} + 15 g_0^3 \int \frac{\mathrm{d}^4 q}{(2\pi)^4}\frac{1}{(r_0 + q^2)^3}. \tag{12.4}$$

In the same spirit, we want to keep the same physics, namely the same G_6, when rescaling Λ. Defining

$$\beta_6 = \Lambda \left.\frac{\partial}{\partial \Lambda} g_6 \right|_{G_6}, \tag{12.5}$$

one finds, dropping terms of higher degree in g_0 (which goes to zero in the scaling limit),

$$\beta_6 = 2g_6 + \frac{15}{4} g_0 g_6 + \cdots . \tag{12.6}$$

In dimension $d \leq 4$, to lowest order in $\epsilon = 4 - d$, this becomes

$$\beta_6 = (2d - 6)g_6 + \frac{15}{4} g_0 g_6 + \cdots . \tag{12.7}$$

Given that $g_0(\lambda) \sim 1/\log \lambda$, this shows that

$$g_6(\lambda) \sim \lambda^{2d-6} \tag{12.8}$$

when $\lambda \to 0$: it is, indeed, irrelevant.

The shift of g_0

The g_6 term also has the effect of 'feeding' in the φ^4 coupling constant, since we now obtain from (12.3)

$$\begin{aligned} g &= \left.\frac{\partial^4 \gamma(m)}{\partial m^4}\right|_{m=0} \\ &= g_0 - \frac{3g_0^2}{2} \int \frac{\mathrm{d}^4 q}{(2\pi)^4} \frac{1}{(q^2 + r_0)^2} + \frac{g_6}{2\Lambda^2} \int \frac{\mathrm{d}^4 q}{(2\pi)^4} \frac{1}{(q^2 + r_0)} . \end{aligned} \tag{12.9}$$

In the large Λ-limit, the last integral diverges as Λ^2, and with the coefficient in front it has a finite limit, some number that adds to g_0, that we could incorporate by shifting ab initio g_0 to g_1, but it is of course simpler to drop g_6 altogether.

In conclusion, there is no reason to assume that the short-distance microscopic Hamiltonian is a renormalizable φ^4 theory. However, the long-distance limit at criticality is effectively given by the renormalizable φ^4 theory.

This analysis applies to high-energy physics as well. Once the theory is regularized by a short-distance cut-off, the field derivatives are in effect replaced by a difference of the field values at nearby sites; the physical mass is much smaller than the cut-off, thus the Green functions decay with distance with a characteristic length m^{-1} much larger than the distance between neighbouring sites. In that sense, a renormalized theory exists only at a critical point. The previous analysis applies *mutatis mutandis*: at very high energy, or very short distances, the physics is still unknown. The fact that quantum electrodynamics is described by a renormalizable theory (minimal coupling) does not come from the unphysical assumption that the theory is valid down to vanishingly small length scales. On the contrary, renormalizability results from the fact that we are considering energies much smaller than the cut-off, at which some other still unknown physics takes place. In that sense, quantum electrodynamics, and, more, generally renormalizable quantum field theories, are simply *effective* critical theories.

Finally, let us recapitulate all this in the light of the general renormalization group picture of Section 10.4. We started with an arbitrary bare theory, constrained simply by symmetry. There is a fixed point at which all the non-renormalizable interactions vanish. Those non-renormalizable terms, if included, flow to the fixed point: they are irrelevant. The φ^4 coupling constant g_0 is also irrelevant in $(4-\epsilon)$ dimension, but the fixed point is non-trivial; it is non-perturbative, except if we perform an expansion in powers of ϵ. In four dimensions, g_0 is a priori marginal; the fixed point is $g_0^\star = 0$ and the flow around the fixed point is non-linear. A one-loop calculation reveals that, under rescaling, g_0 flows to the trivial Gaussian fixed point at the origin. However, the non-linear nature of the flow equation implies that $g_0(\lambda)$ goes to zero only as $1/\log\lambda$, implying logarithmic corrections to the mean-field behaviour.

Clearly, we are assuming here that all the initial coupling constants lie in the basin of attraction of the fixed point. Could there exist another infrared-stable fixed point for g_0? Obviously, this is a non-perturbative question, and the belief in the non-existence of another fixed point comes from the evidence that it has never been found in numerical simulations. This is often called *the triviality of φ^4 in four dimensions.*[3] The consequence is that, if we tried to define a renormalized theory φ^4 valid down to vanishing small distances, we would have to take a theory with a vanishing coupling constant. This 'triviality' is welcome for critical phenomena, but not acceptable for a field theory of elementary physics, which would pretend to describe physics at all length scales. The only escape there lies in a theory in which the fixed point at zero coupling is not an infrared attractor. As is well-known in non-Abelian theories, the origin in coupling constant space is an ultraviolet attractor,[4] a property called *asymptotic freedom*, and they are the only theories endowed with this property.[5]

12.2 Renormalization of the massless theory

We consider the massless, or critical, φ^4 theory

$$A\{\varphi_0\} = \int \mathrm{d}^d x \left\{ \frac{1}{2}(\vec{\nabla}\varphi_0)^2 + \frac{1}{2} r_{0c}{\varphi_0}^2 + \frac{g_0}{4!}{\varphi_0}^4 \right\}, \tag{12.10}$$

in which the relevant coupling constant r_0 has been chosen at its critical value r_{0c}, at which the correlation length is infinite (i.e., the mass vanishes). The theory is

[3] See K. Wilson and J. Kogut, The renormalization group, *Phys. Rep.*, **12c** (1974) 75.
[4] D. J. Gross and F. Wilczek, *Phys. Rev. Lett.*, **30** (1973) 1343. H. D. Politzer, *Phys. Rev. Lett.*, **30** (1973) 1346.
[5] S. Coleman and D. J. Gross, The price of asymptotic freedom, *Phys. Rev. Lett.*, **31** (1973) 851.

regularized by considering its extension in dimension $d = 4 - \epsilon$. In units in which [Length] $= -1$, one has $[\varphi_0 = (d-2)/2]$, $[r_{0c}] = 2$, $[g_0] = \epsilon$, so that $[A = 0]$. This theory is renormalized through

- A field renormalization $\varphi_0 = \sqrt{Z}\varphi$.
- A mass renormalization $r_0 \to m^2$, but here for $r_0 = r_{0c}$, $m = 0$.
- A coupling constant renormalization $g_0 Z^2 = \mu^\epsilon g Z_1$, in which an arbitrary scale factor μ with dimension one has been inserted to make the renormalized coupling constant g dimensionless. We thus have the action in terms of renormalized parameters

$$A\{\varphi\} = \int \mathrm{d}^d x \left\{ \frac{1}{2} Z(\vec{\nabla}\varphi)^2 + \frac{1}{4!}\mu^\epsilon g Z_1 \varphi^4 \right\}. \tag{12.11}$$

The renormalization constants Z and Z_1 are dimensionless power series in g fixed by the renormalization conditions (i.e., the finiteness of the correlation functions). Their series expansions are characterized by coefficients which diverge when $\epsilon \to 0$:

$$Z = 1 + \sum_{k=2}^{\infty} g^k a_k(\epsilon)$$
$$Z_1 = 1 + \sum_{k=1}^{\infty} g^k b_k(\epsilon). \tag{12.12}$$

The *minimal subtraction scheme* is defined by the fact that $a_k(\epsilon)$ and $b_k(\epsilon)$ have poles in ϵ and no finite part

$$a_2(\epsilon) = \frac{A_2}{\epsilon}, \quad a_3(\epsilon) = \frac{A_3^{(1)}}{\epsilon^2} + \frac{A_3^{(2)}}{\epsilon}, \quad \dots$$
$$b_1(\epsilon) = \frac{B_1}{\epsilon}, \quad b_2(\epsilon) = \frac{B_2^{(1)}}{\epsilon^2} + \frac{B_2^{(2)}}{\epsilon}, \quad \dots . \tag{12.13}$$

The renormalization group equations for the coupling constant g are obtained by varying μ at fixed g_0, since the parameter μ is absent from the bare (i.e., unrenormalized) theory:

$$\mu \frac{\partial}{\partial \mu} g \bigg|_{g_0} = \beta(g) \tag{12.14}$$

and since $g_0 = \mu^\epsilon g Z_1 / Z^2$ this gives

$$0 = \epsilon + \beta(g) \frac{\mathrm{d}}{\mathrm{d}g} \log \frac{g Z_1(g)}{Z^2(g)}, \tag{12.15}$$

i.e.,

$$\beta(g) = -\frac{\epsilon g}{1 + g \frac{\mathrm{d}}{\mathrm{d}g} \log \frac{Z_1(g)}{Z^2(g)}}. \tag{12.16}$$

If we want to compute the β function up to order g^3, we take the above expansions (12.12):

$$\frac{Z_1}{Z^2} = \frac{1 + gb_1 + g^2 b_2 + O(g^3)}{1 + 2g^2 a_2 + O(g^3)}$$
$$= 1 + gb_1 + g^2(b_2 - 2a_2) + O(g^3), \qquad (12.17)$$

from which one finds easily

$$\left(1 + g\frac{\mathrm{d}}{\mathrm{d}g}\log\frac{Z_1(g)}{Z^2(g)}\right)^{-1} = 1 - gb_1 + g^2(4a_2 + 2b_1^2 - 2b_2) + O(g^3) \qquad (12.18)$$

and thus

$$\beta(g) = -\epsilon g + \epsilon b_1 g^2 - \epsilon(4a_2 + 2b_1^2 - 2b_2)g^3 + O(g^4), \qquad (12.19)$$

replacing the a_k and b_k by their expressions (12.13)

$$\beta(g) = -\epsilon g + B_1 g^2 + O(g^3). \qquad (12.20)$$

However, at next order, the finiteness of the renormalized theory, and in particular of $\beta(g)$, requires the consistency condition:

$$B_1^2 = 2B_2^{(1)}, \qquad (12.21)$$

and then

$$\beta(g) = -\epsilon g + B_1 g^2 + (2B_2^{(2)} - 4A_1)g^3 + O(g^4). \qquad (12.22)$$

It remains thus to compute the coefficients A and B. Note that in this minimal subtraction scheme the β function in dimension d is completely determined by the β function in dimension 4

$$\beta_d(g) = \beta_4(g) - \epsilon g. \qquad (12.23)$$

12.3 The renormalized critical free energy (at one-loop order)

We follow again the background field technique of Chapter 9, giving the renormalized free energy; at one-loop order, it is

$$\Gamma^{(1)}(\varphi) = A(\varphi) + \frac{1}{2}\mathrm{Tr}\log K, \qquad (12.24)$$

with

$$K(x, y) = \delta(x - y)\left(-Z\nabla^2 + \frac{1}{2}\mu^\epsilon g Z_1 \varphi^2\right). \qquad (12.25)$$

The renormalization constants Z and Z_1 at one-loop, i.e., first order in the coupling constant g, are such that $\Gamma^{(1)}(\varphi)$ is finite (when ϵ goes to zero) for any field $\varphi(x)$.

Since we limit ourselves here to a one-loop calculation, we can replace Z and Z_1 by one in K, neglecting terms of order g^2. The coefficient of φ^4 in $\Gamma^{(1)}$ is

$$\frac{1}{4!}\mu^\epsilon g Z_1 \int \mathrm{d}^d x \varphi^4(x) - \frac{1}{16}\int \mathrm{d}^d x \mathrm{d}^d y \left\langle x \left| \frac{1}{-\nabla^2} \right| y \right\rangle \left\langle y \left| \frac{1}{-\nabla^2} \right| x \right\rangle \varphi^2(x)\varphi^2(y), \tag{12.26}$$

and, in Fourier space,

$$\begin{aligned}\left\langle x \left| \frac{1}{-\nabla^2} \right| y \right\rangle &= \int \frac{\mathrm{d}^d p}{(2\pi)^d} \frac{\mathrm{e}^{\mathrm{i}\mathbf{p}(\mathbf{x}-\mathbf{y})}}{p^2} \\ &= \frac{2^{d/2-2}}{(2\pi)^{d/2}} \Gamma(d/2-1) \frac{1}{|\mathbf{x}-\mathbf{y}|^{d-2}}.\end{aligned} \tag{12.27}$$

Proof of (12.27)

$$\int \frac{\mathrm{d}^d p}{(2\pi)^d} \frac{\mathrm{e}^{\mathrm{i}\mathbf{p}\mathbf{r}}}{p^2} = \int_0^\infty \mathrm{d}t \int \frac{\mathrm{d}^d p}{(2\pi)^d} \mathrm{e}^{\mathrm{i}\mathbf{p}\mathbf{r}} \mathrm{e}^{-tp^2} = \frac{\pi^{d/2}}{(2\pi)^d} \int_0^\infty \mathrm{d}t\, t^{-d/2} \mathrm{e}^{-r^2/(4t)}.$$

The change $t \to r^2/(4t)$ readily yields (12.27).

Then

$$\left\langle x \left| \frac{1}{-\nabla^2} \right| y \right\rangle \left\langle y \left| \frac{1}{-\nabla^2} \right| x \right\rangle = \frac{2^{d-4}}{(2\pi)^d} \Gamma^2(d/2-1) \frac{1}{r^{(2d-4)}},$$

where $\vec{r} = \vec{x} - \vec{y}$ and

$$\frac{1}{r^{(2d-4)}} = \frac{1}{\epsilon} S_d \delta(\vec{r}) + \text{f.p.}, \tag{12.28}$$

where the second term is finite when $\epsilon = 4 - d \to 0$, and $S_d = 2\pi^{d/2}/\Gamma(d/2)$ is the surface of the unit sphere.

Proof of (12.28)

Let $f(r)$ be a test function (rapidly decreasing at infinity)

$$\begin{aligned}\int \mathrm{d}^d r \frac{f(r)}{r^{2d-4}} &= S_d \int_0^\infty \mathrm{d}r\, r^{\epsilon-1} f(r) \\ &= -\frac{S_d}{\epsilon} \int_0^\infty \mathrm{d}r\, r^\epsilon f'(r) = \frac{S_d}{\epsilon} f(0) + O(\epsilon^0).\end{aligned}$$

Returning to (12.26) we see that the quartic terms, including the one-loop corrections, are finite for any $\varphi(x)$, provided we take

$$Z_1 = 1 + \frac{3}{2} K_d \frac{g}{\epsilon} + O(g^2) \tag{12.29}$$

and

$$Z = 1 + O(g^2), \tag{12.30}$$

from which we obtain, as explained earlier, $B_1 = 3/2$ and the one-loop β function.

Exercise

Repeat the same calculation for an $O(n)$ *symmetry and show that the factor* $\frac{3}{2}$ *above is replaced by* $(n+8)/9$.

12.4 Away from T_c

Up to now we have traded the bare coupling constant g_0 for the renormalized one g and renormalized the field $\varphi_0 \to \varphi = Z^{-1/2}\varphi_0$. Away from the critical point, one more renormalization is needed, that of the energy operator φ^2, which is not equal to $(\varphi)^2$. Indeed, consider the massive action

$$A\{\varphi\} = \int \mathrm{d}^d x \left\{ \frac{1}{2} Z(\vec{\nabla}\varphi)^2 + \frac{1}{2} Z_2 t \varphi^2 + \frac{1}{4!}\mu^\epsilon g Z_1 \varphi^4 \right\}, \tag{12.31}$$

in which $t \simeq (T - T_c)$ and Z_2 is a power series in g,

$$Z_2 = 1 + \sum_{k=1}^{\infty} g^k c_k(\epsilon), \tag{12.32}$$

determined by the finiteness of the renormalized free energy when $\epsilon \to 0$. Then, at one-loop order,

$$\Gamma^{(1)}(\varphi) = \int \mathrm{d}^d x \left\{ \frac{1}{2} Z(\vec{\nabla}\varphi)^2 + \frac{1}{2} Z_2 t \varphi^2 + \frac{1}{4!}\mu^\epsilon g Z_1 \varphi^4 \right\} + \frac{1}{2}\mathrm{Tr}\log K, \tag{12.33}$$

with

$$K(x,y) = \delta(x-y)\left(-Z\nabla^2 + tZ_2 + \frac{1}{2}\mu^\epsilon g Z_1 \varphi^2\right). \tag{12.34}$$

Again, at one-loop order we can replace Z, Z_1, Z_2 by one in the kernel $K(x,y)$. The coefficient of $\varphi^2(x)$ in $\Gamma^{(1)}$ is then

$$\frac{1}{2} Z_2 t \varphi^2(x) + \frac{1}{4}\mu^\epsilon g \left\langle x \left| \frac{1}{-\nabla^2 + t} \right| x \right\rangle \varphi^2(x).$$

Since

$$\begin{aligned}\left\langle x \left| \frac{1}{-\nabla^2 + t} \right| x \right\rangle &= \int \frac{\mathrm{d}^d p}{(2\pi)^d} \frac{1}{p^2 + t} \\ &= -\frac{K_d}{2} t^{1-\epsilon/2} \frac{\pi}{\sin \pi\epsilon/2} \\ &= -K_d t/\epsilon + O(\epsilon^0),\end{aligned} \tag{12.35}$$

one sees that, by taking (using again the minimal scheme)

$$Z_2 = 1 + \frac{K_d}{2}\frac{g}{\epsilon} + O(g^2), \tag{12.36}$$

the coefficient of $\varphi^2(x)$ is now finite when $\epsilon \to 0$.

Exercise

Repeat the same calculation for an $O(n)$ *symmetry and show that the factor* $\frac{1}{2}$ *above is replaced by* $(n+2)/6$.

Then we vary μ fixing the coefficients of the bare theory. The coefficient of φ^2_{bare} is tZ_2/Z. If we define the temperature flow as

$$\mu \left.\frac{\partial}{\partial\mu}\right|_{\text{bare}} t = t\kappa(g), \tag{12.37}$$

we obtain

$$\kappa(g) = -\beta(g)\frac{\partial}{\partial g}\log\frac{Z_2}{Z}, \tag{12.38}$$

i.e., at one-loop order

$$\kappa(g) = \frac{n+2}{6}g + O(g^2), \tag{12.39}$$

and we recover the correlation length exponent ν computed in the previous chapter. Note that the calculation of $\kappa(g)$, and consequently of the thermal exponent ν, is obtained through this procedure, in the critical, i.e., massless, theory, in which all the calculations are much simpler.

This introductory chapter is, of course, far from a full account of the theory of renormalization. Many books are available; in particular, in the context of this book, see Zinn-Justin.[6]

[6] J. Zinn-Justin, *Quantum Field Theory and Critical Phenomena*, 3rd edn. (Oxford: Oxford University Press, 2002).

13

Goldstone modes

The existence of 'soft modes', massless excitations in particle physics, governs many properties of the phase in which a continuous symmetry is spontaneously broken. In this chapter, we review first the Goldstone theorem,[1] then the non-linear sigma models, which govern the dynamics of the Goldstone bosons, as well as examples of the physics described by these models.

13.1 Broken symmetries and massless modes

Let $A[\psi]$ be an (Euclidean) action, invariant under a Lie group G. We assume that ψ belongs to some representation of G. One can always assume that G is represented by real orthogonal matrices $D(g)$. (Indeed, since we do not assume any irreducibility, we can always transform a complex unitary representation into a real orthogonal one by splitting real and imaginary parts.)

Under an infinitesimal transformation:

$$\psi_a(x) \to \psi_a(x) + \omega^\alpha t^\alpha_{ab} \psi_b(x) \tag{13.1}$$

(where t^α are generators of G) the invariance of A implies:

$$t^\alpha_{ab} \int \mathrm{d}x \frac{\delta S}{\delta \varphi_a(x)} \psi_b(x) = 0. \tag{13.2}$$

The generators t^α_{ab}, $\alpha = 1, 2, \ldots, N$ are real antisymmetric matrices in this orthogonal representation of G; N is the rank of G.

Example

$$G = O(n), \qquad N = \frac{n(n-1)}{2},$$
$$G = SU(n) \times SU(n), \qquad N = 2(n^2 - 1).$$

[1] J. Goldstone, *Nuovo Cimento*, **19** (1961) 155.

The correlation functions of the theory may all be obtained by the following construction.

Define the partition function

$$Z(h, \varphi) = \int D[\psi] e^{-A(\psi)+\int dx h_a(x)[\psi_a(x)-\varphi_a(x)]} \tag{13.3}$$

(in units in which $\beta = 1$).

Then the free energy is given by $\Gamma(h, \varphi) = -\log Z(h, \varphi)$. Fix $h(\varphi)$ by the constraint:

$$\frac{\delta \Gamma}{\delta h_a(x)} = 0. \tag{13.4}$$

The resulting $\Gamma(\varphi) = \Gamma\{h(\varphi), \varphi\}$ is the free energy as a function of the 'order parameter' φ (which was called $m(x)$ in the previous chapters).

For a uniform $\varphi_a(x) = \varphi$, one remembers that the system is confined in a box of volume V and

$$\Gamma\{\varphi_a(x)\}\Big|_{\varphi_a(x)=\varphi_a} = V\gamma(\varphi_a), \tag{13.5}$$

where $\gamma(\varphi_a)$ is the 'effective potential' of the theory.

At mean field level, when all fluctuations are ignored, one retains only the single configuration $\psi_0(x)$, which maximizes the integrand of (13.3):

$$Z_{cl}(h, \varphi) = e^{-A(\psi_0)+\int dx h(\psi_0-\varphi)}, \tag{13.6}$$

in which $\frac{\delta A}{\delta \psi_0(x)} = h(x)$. Then one finds, in this simple fluctuationless or mean-field approximation,

$$\Gamma_{cl}(h, \varphi) = A(\psi_0) - \int dx h(\psi_0 - \varphi) \tag{13.7}$$

$\delta\Gamma_{cl}/\delta h = 0 \Rightarrow \psi_0 = \varphi$ and, therefore, $\Gamma_{cl}[h(\varphi), \varphi] = A(\varphi)$. If φ is constant in space, $A(\varphi) = \text{Vol} \cdot V(\varphi)$, where $V(\varphi)$ is the part of the action which does not contain derivatives of the field. Finally,

$$\gamma_{cl}(\varphi) = V(\varphi). \tag{13.8}$$

The symmetry is spontaneously broken if the effective potential $\gamma(\varphi)$ has a minimum at non-zero φ. Let v_a be a minimum of $\gamma(\varphi)$, and H be the subgroup of G which leaves v_a invariant:

$$h \in H \leftrightarrow D(h)v_a = v_a. \tag{13.9}$$

The n-point functions of the theory are given by

$$\Gamma^{(n)}_{a_1 \ldots a_n}(x_1 \ldots x_n) = \frac{\delta^n \Gamma(\varphi)}{\delta\varphi_{a_1}(x_1) \ldots \delta\varphi_{a_n}(x_n)}\Bigg|_{\varphi_a = v_a}. \tag{13.10}$$

They are the 'one-particle irreducible' n-point functions for $n > 2$. For $n = 2$, $\Gamma^{(2)}$ is the inverse (in the operator sense) of the correlation function

$$\Gamma^{(2)}_{ab}(x, y) = \left\langle \psi_a(x)\psi_b(y) \right\rangle^{-1}. \tag{13.11}$$

Let us return to the invariance under G. We expect the free energy $\Gamma(\varphi)$ to be invariant under G; it is only because its minima are not G-invariant that the symmetry is broken. Let us prove it.

Consider a variation

$$\varphi_a(x) \to \varphi_a(x) + \omega^\alpha t^\alpha_{ab}\varphi_b(x), \tag{13.12}$$

which gives

$$Z(h, \varphi) \to Z(h, \varphi + \omega^\alpha t^\alpha \varphi). \tag{13.13}$$

In the path integral we perform the corresponding change of variables

$$\psi_a(x) \to \psi_a(x) + \omega^\alpha t^\alpha_{ab}\psi_b(x). \tag{13.14}$$

Note that the Jacobian of this transformation is

$$\det(\delta_{ab} + \omega^\alpha t^\alpha_{ab}) = 1 + \omega^\alpha t^\alpha_{aa} + O(\omega^2) = 1 + O(\omega^2), \tag{13.15}$$

since the generators t^α are antisymmetric (the representation being orthogonal).

Since A is invariant we obtain:

$$Z(h, \varphi + \omega^\alpha t^\alpha \varphi) = \int D[\psi] \mathrm{e}^{-A[\psi] + \int \mathrm{d}x J(\psi - \varphi)} \mathrm{e}^{\int \mathrm{d}x \omega^\alpha t^\alpha_{ab}(\psi_b(x) - \varphi_b(x)) h_a(x)}. \tag{13.16}$$

Expanding both sides to first order in ω^α,

$$Z(h) + \omega^\alpha t^\alpha_{ab} \int \mathrm{d}x \varphi_b(x) \frac{\delta Z}{\delta \varphi_a(x)} = \int D[\psi] \mathrm{e}^{-A[\psi] + \int \mathrm{d}x h(\psi - \varphi)} \times \left(1 + \omega^\alpha \int \mathrm{d}x t^\alpha_{ab} h_a(x)(\psi_b(x) - \varphi_b(x))\right). \tag{13.17}$$

Therefore,

$$t^\alpha_{ab} \int \mathrm{d}x \varphi_b(x) \frac{\delta Z}{\delta \varphi_a(x)} = t^\alpha_{ab} \int \mathrm{d}x h_a(x) \frac{\delta Z}{\delta h_b(x)}. \tag{13.18}$$

Substituting $Z = \mathrm{e}^{-\Gamma}$ and, from the definition (13.4) of $h(\varphi)$: $\frac{\delta Z}{\delta h} = 0$, we find, as expected:

$$t^\alpha_{ab} \int \mathrm{d}x \varphi_b(x) \frac{\delta \Gamma}{\varphi_a(x)} = 0. \tag{13.19}$$

Note that this is identical to the equation, which expressed the invariance of A, as it should be.

Taking the derivative with respect to $\varphi_c(y)$,

$$t^\alpha_{ab}\delta_{bc}\frac{\delta\Gamma}{\delta\varphi_a(y)} + t^\alpha_{ab}\int \mathrm{d}x\varphi_b(x)\frac{\delta^2\Gamma}{\delta\varphi_a(x)\delta\varphi_c(y)} = 0. \tag{13.20}$$

We now set $\varphi_a(x) = v_a$. The first term vanishes by definition, thus:

$$t^\alpha_{ab}\int \mathrm{d}x\, v_b\Gamma^{(2)}_{ac}(x-y) = 0. \tag{13.21}$$

If we now define the Fourier transform

$$\widetilde{\Gamma}^{(2)}_{ac}(p) = \int \mathrm{d}x\mathrm{e}^{\mathrm{i}p(x-y)}\Gamma^{(2)}_{ac}(x-y) \tag{13.22}$$

and the mass matrix

$$(M^2)_{ac} = \widetilde{\Gamma}^{(2)}_{ac}(0), \tag{13.23}$$

the previous relation reads

$$t^\alpha_{ab}v_b(M^2)_{ac} = 0. \tag{13.24}$$

Let us divide the Lie algebra of G into the subalgebra of H and its complement:

$$t^\alpha = \begin{cases} \tau^\alpha, & \alpha = 1\ldots P \quad \text{generators of } H \\ \theta^\alpha, & \alpha = P+1\ldots N \quad \text{remaining generators.} \end{cases} \tag{13.25}$$

Since H is defined by $D(h)v = v$, we have $\tau^\alpha v = 0$. Therefore, the previous identity is empty if t^α belongs to $L(H)$. However for the coset G/H the $(N-P)$ vectors $\theta^\alpha_{ab}v_b$ are non-vanishing, and they are linearly independent. Let

$$u^\alpha_a = \theta^\alpha_{ab}v_b, \qquad \alpha = p+1\ldots n. \tag{13.26}$$

These $(N-P)$ vectors are (left) eigenvectors of $(M^2)_{ab}$ for the eigenvalue zero: there are $(N-P)$ massless modes.

Examples

1. $G = O(n)$, $\varphi_a(x)$ belongs to the vector representation, $a = 1\ldots n$, $N = \frac{n(n-1)}{2}$.
   Assume that the potential has a minimum for some vector $\vec{v}$, H consists of rotations that leave $\vec{v}$ invariant $\Rightarrow$ $H = O(N-1)$. Indeed, take coordinates in which $v = (1, 0, \ldots, 0)$, then the h matrices have the form

$$h = \begin{pmatrix} 1 & 0\ldots0 \\ 0 & \\ \vdots & \Omega \end{pmatrix}. \tag{13.27}$$

   Therefore, $P = \frac{(n-1)(n-2)}{2}$ and the number of Goldstone modes is

$$N - P = \frac{n(n-1)}{2} - \frac{(n-1)(n-2)}{2} = n-1. \tag{13.28}$$

2. $G = SU(n) \times SU(n)$. ψ is a complex $n \times n$ matrix and under an element (U_1, U_2) of G it transforms as $\psi \to U_1 \psi U_2^\dagger$. The action

$$A[\psi] = \int \mathrm{d}x \left\{ \mathrm{Tr}\partial\psi^\dagger\partial\psi - m_0^2 \mathrm{Tr}\psi^\dagger\psi + \frac{1}{2}\lambda_1(\mathrm{Tr}\psi^\dagger\psi)^2 + \frac{1}{2}\lambda_2 \mathrm{Tr}\psi^\dagger\psi\psi^\dagger\psi \right\} \tag{13.29}$$

is indeed invariant under $\psi(x) \to U_1\psi(x)U_2^\dagger$. The minima are given by $\delta A/\delta\psi = 0$ (and adjoint). For constant ψ this gives

$$-m_0^2\psi + \lambda_1\psi \mathrm{Tr}(\psi^\dagger\psi) + \lambda_2\psi\psi^\dagger\psi = 0. \tag{13.30}$$

The solutions are given by $\psi = cU_0$, where $U_0 \in SU(n)$ and c is a number defined by the equation

$$-m_0^2 + \lambda_1|c|^2 n + \lambda_2|c|^2 = 0. \tag{13.31}$$

The subgroup H leaving this minimum invariant consists of pairs (U_1, U_2), such that

$$U_1U_0U_2^\dagger = U_0. \tag{13.32}$$

If $U_1U_0 = V_1$ and $U_0U_2 = V_2$ then $V_1 = V_2$. Therefore, H is isomorphic to the diagonal subgroup (V, V) of G. H thus has rank $(n^2 - 1)$ and G has rank $2(n^2 - 1)$. The number of Goldstone modes in this case is

$$2(n^2 - 1) - (n^2 - 1) = n^2 - 1. \tag{13.33}$$

This model was introduced by Gell-Mann and Lévy[2] as a model in which the lightness of the three pions would be explained by a broken symmetry, in which the pions are Goldstone bosons. For $SU(2) \times SU(2)$ there are indeed three massless modes, and the field ψ is a 2×2 matrix. There is thus a fourth massive particle, which they called the σ-particle, hence the name sigma model (the 'linear' sigma model here, since the group $G = SU(2) \times SU(2)$ acts linearly on the fields).

13.2 Linear and non-linear $O(n)$ sigma models

Let us consider an action invariant under $O(n)$, with the field in a vector representation:

$$A[\psi] = \int \mathrm{d}^d x \left[\partial_\mu\psi^a\partial_\mu\psi^a + V(\vec{\psi}^2) \right], \tag{13.34}$$

in which μ runs from 1 to d and a from 1 to n. Assume that V has the shape of a 'Mexican hat', i.e., a set of minima at $|\vec{\psi}| = 1$. Let $\vec{v}_0$ be a unit vector, corresponding (through boundary conditions, or an infinitesimal external field) to one such minimum. We can then decompose the field $\vec{\psi}(x)$ into

[2] M. Gell-Mann and M. Lévy, *Nuovo Cimento*, **16** (1960) 705.

- A space-dependent rotation of the vector $\vec{v}_0$, which brings it along $\vec{\psi}(x)$, but leaves its unit-norm unchanged,
- Longitudinal excitations $L(x)$, which change the length of $\vec{v}_0$

$$\psi_a(x) = [1 + L(x)]R_{ab}(x)(\vec{v}_0)_b, \tag{13.35}$$

and call $v_a(x)$ the unit vector $R_{ab}(x)(v_0)_b$. In this parametrization,

$$A[L, \vec{v}] = \int \mathrm{d}x \left[(\partial_\mu L)^2 + (\partial_\mu v_a)^2 + (L^2 + 2L)(\partial_\mu v_a)^2 + V((1+L)^2)\right]. \tag{13.36}$$

Expanding the potential about its minimum at $L = 0$, we have $V = V(1) + 2L^2V''(1) + O(L^3)$, $V''(1) > 0$. This action A describes:

- One massive longitudinal excitation $L(x)$ with mass-squared $m^2 = 2V''(1)$,
- $(n-1)$ massless modes v_a, $a = 1, \ldots, n$ constrained by $\vec{v}^2 = 1$.

The properties of this theory at long distance (we are interested in physical distances much larger than the lattice spacing or inverse momentum cut-off, i.e., in the vicinity of a critical point) are governed by diagrams in which only massless v_a particles are exchanged. Any exchange of the massive mode $L(x)$ is subleading in the long distance limit. Therefore, we may drop the longitudinal mode completely from the action and keep only the long-distance Goldstone modes:

$$A = \frac{1}{2\lambda}\int \mathrm{d}x(\partial_\mu v_a)^2, \quad \mu = 1, \ldots, d, \quad a = 1, \ldots, n \tag{13.37}$$

with the non-linear constraint $\vec{v}^2 = 1$. This is the $O(n)$ non-linear sigma model (i.e., the model for the coset $O(n)/O(n-1)$) in which, if we choose a set of $(n-1)$ parameters on the sphere, the $O(n)$-symmetry will be realized non-linearly. For instance, if we choose a parametrization of the northern hemisphere,[3] such as

$$\sigma(x) \equiv v_0(x) = \sqrt{1 - \sum_1^{N-1} v_i^2(x)}, \tag{13.38}$$

a rotation that mixes v_0 and v_i is not a simple linear operation on the independent variables v_a, $a = 1, \ldots, n-1$.

The regularization of this theory by a lattice reveals that this model is nothing but the familiar classical n-vector model of statistical physics. Indeed, on a lattice there is a unit vector $v_a(x_i)$ at each lattice site; the derivative $\partial_\mu v_b$ is replaced by

[3] As in any saddle-point expansion the error in choosing this parametrization, which singles out one hemisphere, is of order $\exp -1/\lambda$.

$(v_b(x+ae_\mu)-v_b(x))/a$, where a denotes the lattice spacing and e_μ the unit vector in direction μ. Hence,

$$\sum_{b=1}^{n}(\partial_\mu v_b)^2 \to -\frac{2}{a^2}\vec{v}(x)\cdot\vec{v}(x+ae_\mu)+\text{const.} \tag{13.39}$$

We see that the non-linear sigma model on the lattice is nothing but the n-vector model with the Hamiltonian $H=-J\sum_{\langle ij\rangle}\vec{v}_i\cdot\vec{v}_j,\ \vec{v}^2=1$.

In statistical mechanics the coupling constant λ is proportional to the absolute temperature (given that the Boltzmann weight is $\mathrm{e}^{-H/T}$). The non-linear sigma model with coupling constant λ is thus the long distance limit of the n-vector model at temperature λ (in the appropriate units). We have already encountered this model in Section 5.3, in which we were performing a low-temperature expansion for the n-vector model.[4] The sigma models have played an important rôle in many areas of physics: expansion near two dimensions of critical quantities, Anderson localization, disordered systems. They are the starting point of string theory in which the field is a mapping of a two-dimensional world-sheet into a space–time manifold.

13.3 Regularization and renormalization of the $O(n)$ non-linear sigma model in two dimensions

13.3.1 Regularization

The problem has both infrared and ultraviolet singularities, which need to be regularized.

As far as the infrared singularities are concerned, the problem is not too severe. A finite box, a mass term or more conveniently, as we shall see, an external field, will take care of this problem. The situation is more delicate with the ultraviolet singularities.

Indeed, let us write the model using some parametrization of the sphere $\vec{v}^2=1$, for instance

$$\vec{v}=(\sqrt{1-\vec{\varphi}^2},\varphi_1,\ldots,\varphi_{n-1}). \tag{13.40}$$

Then

$$(\partial_\mu\vec{v})^2=g_{ij}(\varphi)\partial_\mu\varphi_i\partial_\mu\varphi_j, \tag{13.41}$$

[4] This model was introduced in A. Polyakov, Interaction of Goldstone Particles in two dimensions: applications to ferromagnets and massive Yang–Mills fields, *Phys. Lett. B*, **59** (1975) 79–81. Performing a one-loop calculation, he showed that the model was 'asymptotically free'. The renormalizability to all orders was proven in E. Brézin and J. Zinn-Justin, Renormalisation of the non-linear sigma-model in $2+\epsilon$ dimensions; application to the Heisenberg ferromagnets, *Phys. Rev. Lett.*, **361** (1976) 69 and *Phys. Rev. D*, **14** (1976) 2615 (with J. C. Le Guillou).

with

$$g_{ij} = \delta_{ij} + \frac{\varphi_i \varphi_j}{1 - \vec{\varphi}^2}, \tag{13.42}$$

in which φ^2 stands for $\varphi_1^2 + \cdots + \varphi_{n-1}^2$. The field φ_i is dimensionless in two dimensions. Therefore, the renormalized action will contain every possible term of dimension two, not protected by symmetry, thus a priori any function $g_{ij}^R(\varphi)$ multiplied by $\partial_\mu \varphi_i \partial_\mu \varphi_j$. We could thus generate an arbitrary metric on the manifold, and indeed this is what happens with a sigma model on an arbitrary Riemannian manifold.

In the parametrization that we have chosen, the $O(n-1)$ symmetry consisting of rotations $\varphi_i \to R_{ij}\varphi_j$ is manifest since under that rotation $g_{ij}\partial_\mu \varphi_i \partial_\mu \varphi_j$ remains invariant. However, rotations that mix the $(1, 0, \ldots, 0)$ vector with the other directions are more subtle. For example, one can verify that, as expected, the action is invariant under $\vec{\varphi} \to \vec{\varphi} + \delta\vec{\varphi}$ with, for one of the components of φ, say φ_1,

$$\delta\varphi_1 = \epsilon\sqrt{1 - \vec{\varphi}^2}, \qquad \delta\sqrt{1 - \vec{\varphi}^2} = -\epsilon\varphi_1. \tag{13.43}$$

A cut-off on the $(n-1)$ components $\varphi_1 \ldots \varphi_{n-1}$ will, in general, break this full non-linear $O(n)$ symmetry and counter-terms could proliferate. (For instance one can check that if the propagator of the φ_i is cut off to zero above some momentum Λ, the correlation function of the longitudinal mode $\sqrt{1 - \vec{\varphi}^2}$ has a support that extends beyond Λ.)

There are two schemes that are satisfactory in this respect, since they manifestly preserve the $O(n)$ symmetry:

Lattice regularization

One returns to the lattice n-vector model with its manifest $O(n)$ symmetry,

$$H = J \sum_{ij} (\vec{S}_i - \vec{S}_j)^2, \tag{13.44}$$

i.e., with the same parametrization of the unit sphere:

$$(\vec{S}_i - \vec{S}_j)^2 = (\vec{\varphi}_i - \vec{\varphi}_j)^2 + \left(\sqrt{1 - \vec{\varphi}_i^2} - \sqrt{1 - \vec{\varphi}_j^2}\right)^2. \tag{13.45}$$

The quadratic form in φ_i in this action is simply $(\vec{\varphi}_i - \vec{\varphi}_j)^2$, giving a propagator in momentum space:

$$\frac{T}{J\sum_{\mu=1}^{d}(1 - \cos p_\mu)}. \tag{13.46}$$

The interaction is given by expanding $\sqrt{1-\vec{\varphi}_i^2}$ in powers of φ_i. One should also keep in mind the invariant measure over the unit sphere:

$$Z = \int \prod_i \frac{d\vec{\varphi}_i}{\sqrt{1-\vec{\varphi}_i^2}} e^{-H/T}. \tag{13.47}$$

The measure terms must be taken into account as further interactions:

$$e^{-H/T} \to e^{-H/T - \frac{1}{2}\sum_i \log(1-\vec{\varphi}_i^2)}.$$

This procedure is safe and well-defined once the infrared problem is taken care of. However, lattice propagators make calculations, beyond the simple one-loop order, rather painful.

Dimensional regularization

This is, of course, considerably easier. By going to dimension $d < 2$ all ultraviolet divergences are well-regulated (and in particular the measure terms may be dropped as a consequence of the rules of d-dimensional intergration). However, the infrared problem becomes worse. We have thus to take care of the infrared divergences before considering dimensional regularization.

We could simply add a mass term to the action:

$$A = \frac{1}{2\lambda} \int d^d x \left[(\partial_\mu \vec{v})^2 + \frac{m^2}{2} \vec{\varphi}^2 \right], \tag{13.48}$$

breaking of course the full $O(n)$ symmetry, but with a 'soft operator', which will not affect the ultraviolet singularities. This is perfectly feasible but implies discussing the renormalization of $\vec{\varphi}^2$ in addition to that of $\vec{\varphi}$. Therefore, we choose a more complicated 'soft' operator, which, thanks to the $O(n)$ symmetry, turns out to be simpler. Consider:

$$A = \int d^d x \left[\frac{1}{2\lambda} (\partial_\mu \vec{v})^2 - \frac{h}{\lambda} \sqrt{1-\vec{\varphi}^2} \right], \tag{13.49}$$

which in spin language means that we have added a magnetic field h along the $(1, 0, \ldots, 0)$ direction, which couples as $\vec{h} \cdot \sum_i \vec{S}_i$. The quadratic form of this action is now

$$A^0 = \int d^d x \left[\sum_{i=1}^{n-1} \left(\frac{1}{2\lambda} (\partial_\mu \varphi_i)(\partial_\mu \varphi_i) + \frac{h}{2\lambda} \varphi_i \varphi_i \right) \right]. \tag{13.50}$$

The $(n-1)$ Goldstone bosons now have a mass-squared h, since the propagator is

$$\langle \varphi_i \varphi_j \rangle = \frac{\lambda}{p^2 + h} \delta_{ij}. \tag{13.51}$$

Note that this regulator also has the effect of adding new vertices to the interaction, since

$$\sqrt{1-\vec{\varphi}^2} = 1 - \frac{\vec{\varphi}^2}{2} - \frac{1}{8}(\vec{\varphi}^2)^2 + \cdots . \tag{13.52}$$

However, because of the full $O(n)$ symmetry, which is only softly broken, we expect that if φ_i is renormalized multiplicatively as $\varphi_i \to \sqrt{Z}\varphi_i^R$, the composite operator $\sqrt{1-\vec{\varphi}^2}$ is similarly renormalized by the same factor $\sqrt{Z}$. Therefore, the field-strength renormalization is sufficient to renormalize the infrared regulator term. This is why we prefer it to a straight mass term, which would involve a distinct renormalization constant.

Now that the infrared problem is under control, we may regularize the ultraviolet singularities by going to dimension $d < 2$. We then choose to renormalize the theory through the minimal subtraction scheme, in which the renormalization constants contain only poles in $(d-2)$ (and no finite part). Finally, after all these poles have been removed by minimal subtraction, we obtain a theory that is finite, order by order in a double expansion in powers of the temperature and of $(d-2)$.

Let us finally note that dimensional regularization makes the measure terms, which are essential in the lattice regularization, very easy to handle. Indeed,

$$\prod_i \frac{\mathrm{d}\vec{\varphi}_i}{\sqrt{1-\vec{\varphi}^2}} = \prod_i \mathrm{d}\vec{\varphi}_i \, \mathrm{e}^{\left[-\frac{1}{2}\sum_i \log(1-(\vec{\varphi}_i)^2)\right]} \tag{13.53}$$

flows formally in the continuum limit to

$$\prod_x \mathrm{d}\vec{\varphi}(x) \, \mathrm{e}^{\left[-\frac{1}{2a^d}\int \mathrm{d}^d x \log(1-\vec{\varphi}^2(x))\right]}. \tag{13.54}$$

But $(2\pi)^d/a^d = \int_{\mathrm{BZ}} \mathrm{d}^d p$, where BZ is the Brillouin zone, in which each component of p varies between $(-\pi/a, \pi/a)$. In the continuum limit,

$$\frac{1}{a^d} = \int \frac{\mathrm{d}^d p}{(2\pi)^d} \tag{13.55}$$

is just the integral of one which vanishes, by definition, since in dimensional regularization

$$\int \frac{\mathrm{d}^d p}{(2\pi)^d} \frac{1}{(p^2)^k} = 0, \quad \forall k. \tag{13.56}$$

Therefore, the measure terms may be ignored altogether in this regularization.

The strategy exposed here is thus (i) to eliminate infrared singularities by the action of an external magnetic field, (ii) continue to dimension $d < 2$ to regularize the ultraviolet divergences, (iii) by minimal subtraction, construct a well-defined renormalized theory in dimension two and (iv) continue if necessary to $d > 2$, to derive an expansion for the critical properties in powers of $(d-2)$ similar, for continuous symmetries, to the earlier expansion in powers of $(4-d)$.

13.3.2 Perturbation expansion and renormalization

We now collect the non-Gaussian terms of the action

$$A\{\varphi\} = \frac{1}{\lambda}\int \mathrm{d}^d x \left[\frac{1}{2} g_{ij}(\varphi)\partial_\mu\varphi_i\partial_\mu\varphi_j - h\sqrt{1-\vec{\varphi}_i^2}\right], \tag{13.57}$$

with

$$g_{ij}(\varphi) = \delta_{ij} + \frac{\varphi_i\varphi_j}{1-\vec{\varphi}^2}. \tag{13.58}$$

The part of this action that is quadratic in the field is

$$A^{(0)}\{\varphi\} = \frac{1}{2\lambda}\int \mathrm{d}^d x[\partial_\mu\vec{\varphi}\partial_\mu\vec{\varphi} + h\vec{\varphi}^2], \tag{13.59}$$

giving a propagator in momentum space equal to $\frac{\lambda}{p^2+h}$. The quartic part in this action is

$$A^{(1)}\{\varphi\} = \frac{1}{8\lambda}\int \mathrm{d}^d x[\partial_\mu(\vec{\varphi})^2\partial_\mu(\vec{\varphi})^2 + h(\vec{\varphi}^2)^2]. \tag{13.60}$$

However, there is an infinite number of higher vertices of the form

$$\frac{1}{\lambda}(\vec{\varphi}^2)^k\varphi^i\partial_\mu\varphi^j\varphi^i\partial_\mu\varphi^j$$

for those coming from the metric term; they contain $(2k+4)$ lines and two derivatives. Those coming from the h terms contain $\frac{h}{\lambda}(\vec{\varphi}^2)^k$ with some binomial coefficient. An arbitrary Feynman diagram with V vertices and I internal lines carries a factor

λ^{-V}, since each interaction vertex carries a factor $1/\lambda$,

λ^{I}, since the propagator is $\dfrac{\lambda}{h+p^2}$,

but $I-V=L-1$, where L is the number of loops of the diagram. Therefore, the loop expansion is an expansion in powers of λ, i.e., a low-temperature expansion.

To give an idea of the renormalization in this model, let us consider at one loop the inverse propagator

$$\Gamma^{(2)}_{ij}(p) = \delta_{ij}\frac{1}{\lambda}[p^2+h] + (\text{Diag. 1}) + (\text{Diag. 2}) + O(\lambda). \tag{13.61}$$

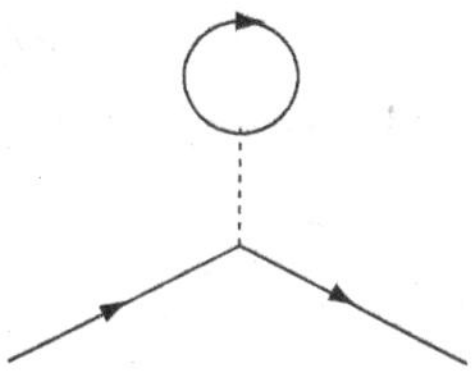

Diag. $1 = \frac{1}{2}(n-1)h\int\frac{d^dq}{q^2+h}$ Diag. $2 = \int\frac{d^dq(q^2+h)}{(p-q)^2+h}$

Using dimensional regularization, continuing to dimension $d = 2-\epsilon$, one finds:

$$\Gamma^{(2)}_{ij}(p) = \delta_{ij}\frac{1}{\lambda}(p^2+h) + K_d\delta_{ij}\left[\frac{h^{-\epsilon/2}}{\epsilon}p^2 + \frac{n-1}{2}\frac{h^{1-\epsilon/2}}{\epsilon} + \text{finite terms}\right] + O(\lambda), \quad (13.62)$$

where $K_d \equiv \frac{2\pi^{d/2}}{\Gamma(d/2)(2\pi)^d}$.

The finiteness of $\Gamma^{(2)}$ is achieved through field amplitude and coupling-constant renormalizations. If we define $\varphi^a = \sqrt{Z}\varphi^a_{\text{R}}$, $a = 1,\ldots,n-1$ and $\sigma = \sqrt{1-\vec{\varphi}^2} = \sqrt{Z}\sigma_{\text{R}}$, Z will be fixed by the requirement that φ_{R} and σ_{R} have finite correlation functions in two dimensions. Note that $\sigma_{\text{R}} = \frac{1}{\sqrt{Z}}\sqrt{1-Z\vec{\varphi}_{\text{R}}^{\,2}}$ is a finite composite operator (since we have kept the $O(n)$ symmetry) and it does not carry an anomalous dimension on its own. The coupling-constant renormalization is defined as $\lambda = \lambda_{\text{R}}Z_1\mu^{2-d}$, in which μ is an arbitrary momentum scale that makes λ_{R} dimensionless. (The $O(n)$ symmetry does not allow for more than one coupling constant.)

The renormalizability of this theory is the statement that Z and Z_1 are power series in λ_{R}, whose coefficients have poles when $d-2 \to 0$, such that the correlation functions of φ_{R}, σ_{R}, expressed as power series in λ_{R}, are finite.

The field 'renormalizes' into a field that couples to σ_{R}, which yields finite expectation values

$$\frac{h}{\lambda}\sqrt{1-\vec{\varphi}^2} \to \frac{h_{\text{R}}}{\lambda_{\text{R}}}\mu^{d-2}\sigma_{\text{R}}. \quad (13.63)$$

Therefore,

$$\sqrt{Z}\frac{h}{Z_1} = h_{\text{R}}, \quad (13.64)$$

and the action in terms of renormalized fields reads

$$A = \frac{\mu^{d-2}}{2Z_1\lambda_R}\int d^d x\left[Z(\partial_\mu\vec{\varphi}_R)^2 + \left(\partial_\mu\sqrt{1 - Z\vec{\varphi}_R^{\,2}}\right)^2\right] - \frac{h_R}{\lambda_R}\frac{\mu^{d-2}}{\sqrt{Z}}\int d^d x\sqrt{1 - Z\vec{\varphi}_R^{\,2}}. \tag{13.65}$$

The quadratic part is simply

$$\frac{\mu^{d-2}}{2\lambda_R}\int d^d x\left[\frac{Z}{Z_1}(\nabla\varphi_R)^2 + h_R\sqrt{Z}\varphi_R^2\right]. \tag{13.66}$$

Returning to the previous calculation we see that it fixes Z/Z_1 and $\sqrt{Z}$ to be at one-loop order:

$$\begin{aligned}\frac{Z}{Z_1} &= 1 - \frac{1}{\epsilon}\lambda_R + O(\lambda_R^2),\\ \sqrt{Z} &= 1 - \frac{n-1}{2\epsilon}\lambda_R + O(\lambda_R^2)\end{aligned} \tag{13.67}$$

(we have included in λ_R the trivial factor k_d).

It is straightforward but lengthy to extend this calculation to higher-loop orders.

13.4 Renormalization group equations for the $O(n)$ non-linear sigma model and the $(d-2)$ expansion

The renormalized theory is the scaling limit of the theory, the limit in which the lattice spacing is vanishingly small. The renormalization group equations may be simply derived by expressing that the bare theory does not know the arbitrary length scale μ that we had to introduce in the process of renormalization. Since the irreducible k-point correlation functions satisfy

$$\Gamma_R(p_1\cdots p_k, \lambda_R, h_R, \mu) = Z^{k/2}\Gamma(p_1\cdots p_k, \lambda, h, \Lambda), \tag{13.68}$$

we have

$$\mu\frac{\partial}{\partial\mu}Z^{-k/2}\Gamma_R^{(k)}\Big|_{\text{bare}} = 0. \tag{13.69}$$

Defining

$$\begin{aligned}\mu\frac{\partial}{\partial\mu}\lambda_R\Big|_{\text{bare}} &= \beta[\lambda_R],\\ \mu\frac{\partial}{\partial\mu}\log Z\Big|_{\text{bare}} &= \gamma[\lambda_R],\end{aligned} \tag{13.70}$$

we note that

$$\mu\frac{\partial}{\partial\mu}\lambda_R\Big|_{\text{bare}} = \beta[\lambda_R] = \frac{(d-2)\lambda_R}{1 + \frac{d}{d\lambda_R}\log Z_1} \tag{13.71}$$

(derived from $\lambda = \lambda_R Z_1 \mu^{d-2}$). From $h_R = h\sqrt{Z}/Z_1$, we derive

$$\mu \frac{\partial}{\partial \mu}\bigg|_{\text{bare}} h_R = \left(\frac{1}{2}\gamma(\lambda_R) - (d-2) + \frac{1}{\lambda_R}\beta(\lambda_R)\right) h_R, \tag{13.72}$$

which leads to the differential equation

$$\left[\mu\frac{\partial}{\partial\mu} + \beta\frac{\partial}{\partial\lambda} + \left(\frac{1}{2}\gamma(\lambda) + \frac{\beta(\lambda)}{\lambda} - (d-2)\right) h\frac{\partial}{\partial h} - \frac{1}{2}k\gamma(\lambda)\right]\Gamma^{(k)}(\lambda, h, \mu) = 0, \tag{13.73}$$

in which it is now understood that we deal with the renormalized theory in terms of the renormalized parameters, and thus drop from now on the subscript R. The explicit one-loop result leads to

$$\begin{aligned} \beta(\lambda) &= (d-2)\lambda - (n-2)\lambda^2 + O(\lambda^3), \\ \gamma(\lambda) &= (n-1)\lambda + O(\lambda^2). \end{aligned} \tag{13.74}$$

In the minimal subtraction scheme that we have used, in which Z and Z_1 only have poles in ϵ, it is a standard exercise to prove that to all orders in a double expansion in λ and $(d-2)$ the occurrence of $(d-2)$ takes place only in the first term of the β function, namely the term linear in λ; $(d-2)$ does not appear either in higher-order terms of the β function, or in any term of the function $\gamma(\lambda)$.

In two dimensions, $\beta(\lambda) = -(n-2)\lambda^2 + O(\lambda^3)$: the origin $\lambda = 0$ is an ultraviolet fixed point, the theory is *asymptotically free*; let me stress again that this was expected, since λ, the temperature, is a relevant variable. Calculations along these lines have now been pushed up to four loops.

The case $n = 2$, in which the group of symmetry $O(2)$ is Abelian, is special: the β function vanishes to all orders. This can be seen by the parametrization $v_1 = \cos\theta$, $v_2 = \sin\theta$, with which $(\partial_\mu \vec{v})^2 = (\partial_\mu\theta)^2$, a free field theory ... in perturbation theory in which one forgets that θ is an angle varying on a 2π interval. However, Kosterlitz and Thouless have shown that at low temperatures there are vortices or antivortices, excitations around which the 'spin' makes a complete circle; these excitations unbind at some critical temperature, giving rise to a transition of infinite order.

13.4.1 Integration of RG equations and scaling

Since the equations that we have derived are valid to all orders in λ and $(d-2)$, we can examine their properties for any d. As it was explained, the critical point in a relevant variable such as λ, has to be an ultraviolet fixed point. It is given here in dimension d greater than two by

$$\lambda_c = \frac{d-2}{n-2} + O[(d-2)^2], \tag{13.75}$$

which, as it should, vanishes at $d = 2$.

Any invariant under renormalization group transformation, satisfies the homogeneous renormalization group equation (without the $k\gamma$ term):

$$\left[\mu\frac{\partial}{\partial\mu}+\beta(\lambda)\frac{\partial}{\partial\lambda}\right]\text{Inv.}=0. \tag{13.76}$$

In particular, the mass $m = \xi^{-1}$, or inverse correlation length, satisfies this equation. Given that it has one dimension, this gives

$$m=\mu\exp\left(-\int_{\lambda_0}^{\lambda}\frac{\mathrm{d}\lambda'}{\beta(\lambda')}\right). \tag{13.77}$$

From there we recover:

(i) Dynamical mass generation in two dimensions, since for λ small and $d=2$ we have:

$$m=\mu\mathrm{e}^{-\frac{1}{\lambda(n-2)}}. \tag{13.78}$$

(ii) Above two dimensions, this mass vanishes at λ_c as

$$m=\mu(\lambda_c-\lambda)^{-\frac{1}{\beta'(\lambda_c)}}, \tag{13.79}$$

and at the ultraviolet fixed point $\beta'(\lambda_c)<0$. This yields the correlation length, which diverges at the critical point as

$$\begin{aligned}&\xi\sim(\lambda_c-\lambda)^{-\nu},\\&\nu=-\frac{1}{\beta'(\lambda_c)}=\frac{1}{d-2}+O[(d-2)^2]\end{aligned} \tag{13.80}$$

(the result for two loops is $\frac{1}{\nu}=d-2+\frac{(d-2)^2}{n-2}+O[(d-2)^3]$).

One can also study the spontaneous magnetization $M = \langle\sigma_R\rangle$, which satisfies the $k=1$ equation in zero field. Since $\sigma_R = \sqrt{1-\varphi_R^2}$ is dimensionless, the renormalization group equation reads

$$\left[\beta(\lambda)\frac{\partial}{\partial\lambda}+\frac{1}{2}\gamma(\lambda)\right]M=0, \tag{13.81}$$

from which it follows that

$$M(\lambda)\sim\mathrm{e}^{-\frac{1}{2}\int_0^{\lambda}\mathrm{d}\lambda'\frac{\gamma(\lambda')}{\beta(\lambda')}}. \tag{13.82}$$

This magnetization vanishes at λ_c with the exponent

$$-\frac{\gamma(\lambda_c)}{2\beta'(\lambda_c)}=\frac{n-1}{2(n-2)}+O(d-2).$$

It is then straightforward to verify that the k-point functions satisfy the scaling relation

$$\Gamma^{(k)}(p_1 \dots p_k; \lambda, h = 0, \mu) = m^d(\lambda, \mu)\mu^{-k}(\lambda)F^{(k)}(p_i/m). \tag{13.83}$$

I shall stop here, but one can see how scaling laws, critical exponents, etc., follow from this type of analysis.

13.5 Extensions to other non-linear sigma models

Several fat books would be needed to cover the fields in which non-linear sigma models have been applied to physics. Let me mention here simply two examples.

Sigma model on an arbitrary Riemannian manifold

In the $O(n)$ model the field $\vec{v}(x)$ was a mapping of two-dimensional space into the sphere S^{n-1}. Once we parametrized the sphere by coordinates $\varphi_a(x)$, we obtained the action

$$A(\varphi) = \int \mathrm{d}^2x g_{ab}(\varphi)\partial_\mu\varphi_a\partial_\mu\varphi_b, \tag{13.84}$$

with g_{ab} given by (13.58). This was generalized by Friedan[5] to a mapping from the plane to an arbitrary Riemannian manifold. Then nothing prevents this metric from an arbitrary renormalization, and thus we expect as many renormalization constants as the number of metric tensors on a d-dimensional manifold. We thus expect the renormalization group equations to take a functional form with a β functional

$$\beta_{ab} = \mu\frac{\partial}{\partial\mu}\bigg|_{\text{bare}} g_{ab} \tag{13.85}$$

of all possible metric tensors. The result up to two-loop order is

$$\beta_{ab} = (d-2)g_{ab} - \frac{1}{2\pi}R_{ab} - \frac{1}{8\pi^2}R_{aklm}R_{bklm} + \cdots, \tag{13.86}$$

in which R_{ijkl} is the Riemannian curvature tensor, and $R_{ij} = R^k_{ijk}$, the Ricci tensor. [For symmetric spaces R_{ij} and all the tensors in the r.h.s. are proportional to g_{ij} and one recovers a one-coupling-constant renormalization group equation.]

This result plays an important rôle in string theory. In Polyakov formulation,[6] the string sweeps a two-dimensional manifold in its motion. It is thus described by a mapping of a two-dimensional world-sheet to the manifold. The consistency requires Weyl invariance, i.e., $\beta_{ab} = 0$. The manifold should have vanishing Ricci

[5] D. H. Friedan, *Ann. Phys. (NY)*, **163** (1985) 318.

[6] A. M. Polyakov, *Phys. Lett. B*, **103** (1981) 207.

tensor R_{ij}. Non-trivial manifolds with $R_{ij} = 0$, but the total curvature $R \neq 0$, i.e., 'Ricci-flat', are called Calabi–Yau manifolds.

Localization by a disordered potential

Applications of sigma models to disordered metals or semiconductors are numerous.[7] One of the most important models concerns Anderson localization. Anderson[8] has shown that in one dimension a particle scattered by random impurities does not diffuse: its wave function remains localized, rather than extended over the whole system. Localization in one dimension may be intuitively understood as the results of random traps in the scattering process. More surprisingly, Abrahams and co-workers, in a famous paper,[9] argued that in two dimensions the electrons will also remain localized, at least when the interactions between the electrons are neglected: a two-dimensional disordered metal is thus an insulator. Above two dimensions, the theory, a one-parameter scaling theory, predicts a metal–insulator transition when the disorder gets stronger. This picture was turned into an explicit framework by F. Wegner[10] as follows. Consider a quantum particle subject to a random potential $V(\vec{r})$, the quantum Hamiltonian is

$$H = -\nabla^2 + V(\vec{r}). \tag{13.87}$$

For simplicity, one assumes that this potential is a Gaussian white noise

$$\langle V(\vec{r})\rangle = 0, \qquad \langle V(\vec{r})V(\vec{s})\rangle = w\delta(\vec{r}-\vec{s}). \tag{13.88}$$

The conductivity is related to the average

$$\left\langle < \vec{r}_1 \left|(E + \mathrm{i}\epsilon - H)^{-1}\right| \vec{s}_1 >< \vec{r}_2 \left|(E - \mathrm{i}\epsilon - H)^{-1}\right| \vec{s}_2 > \right\rangle. \tag{13.89}$$

One then uses the representation of the resolvent as a Gaussian integral

$$< \vec{r}_1 \left|(E + \mathrm{i}\epsilon - H)^{-1}\right| \vec{s}_1 >= \frac{N}{D}, \tag{13.90}$$

$$N = \int \mathrm{D}\varphi\, \varphi^\star(\vec{r}_1)\varphi(\vec{s}_1)\mathrm{e}^{\mathrm{i}\int \mathrm{d}\vec{r}\varphi^\star(\vec{r})(E+\mathrm{i}\epsilon-V)\varphi(\vec{r})},$$
$$D = \int \mathrm{D}\varphi \mathrm{e}^{\mathrm{i}\int \mathrm{d}\vec{r}\varphi^\star(\vec{r})(E+\mathrm{i}\epsilon-V)\varphi(\vec{r})}. \tag{13.91}$$

[7] See, e.g., K. Efetov, *Supersymmetry in Disorder and Chaos*, (Cambridge: Cambridge University Press, 1997).
[8] P. W. Anderson. *Phys. Rev.*, **109** (1958) 1492.
[9] E. Abrahams, P. W. Anderson, D. C. Licciardello and T. V. Ramakrishnan, *Phys. Rev. Lett.*, **42** (1979) 673.
[10] F. Wegner, Anderson localization and nonlinear sigma model, *Journ. Stat. Phys.*, **36** (1984) 283.

Averaging a ratio over the randomness is difficult. At this stage, we are left with two choices, replicas or 'supersymmetry'. Since we have not used Grassmannian integrals in this book, I shall sketch here the use of replicas. The identity

$$\frac{N}{D} = \lim_{n\to 0} N D^{n-1} \tag{13.92}$$

leads to

$$\begin{aligned} &< \vec{r}_1 \left|(E + \mathrm{i}\epsilon - H)^{-1}\right| \vec{s}_1 > \\ &= \lim_{n\to 0} \int \prod_{a=1}^{n} \mathrm{D}\varphi_a \varphi_1^\star(\vec{r}_1)\varphi_1(\vec{s}_1)\, \mathrm{e}^{\mathrm{i}\sum_{a=1}^{n}\int \mathrm{d}\vec{r}\varphi_a^\star(\vec{r})(E+\mathrm{i}\epsilon - V)\varphi_a(\vec{r})}. \end{aligned} \tag{13.93}$$

The same is done with $\langle \vec{r}_2|\frac{1}{E-\mathrm{i}\epsilon-H}|\vec{s}_2\rangle$ (but with $\mathrm{i} \to -\mathrm{i}$ in the exponential), through a Gaussian integral over n variables $\psi_a(\vec{r})$. The averaging over the random potential is then immediate since

$$\left\langle \mathrm{e}^{-\mathrm{i}\int \mathrm{d}r V(\vec{r})U(\vec{r})} \right\rangle = \mathrm{e}^{-\int \mathrm{d}r U^2(\vec{r})}, \tag{13.94}$$

with

$$U(\vec{r}) = \sum_{1}^{n} [\varphi_a^\star(\vec{r})\varphi_a(\vec{r}) - \psi_a^\star(\vec{r})\psi_a(\vec{r})]. \tag{13.95}$$

The resulting effective action has a pseudo-orthogonal symmetry $O(n, n)$ (i.e., rotations in $2n$ dimensions preserve the quadratic form $x_1^2 + \cdots + x_n^2 - x_{n+1}^2 - \cdots - x_{2n}^2$), broken only infinitesimally by the $\mathrm{i}\epsilon$ term to the subgroup $O(n) \times O(n)$. The non-linear sigma model for the 'Grassmannian' manifold $O(n, n)/O(n) \times O(n)$ is easy to formulate.[11] The calculation of the β function confirms the analysis of Abrahams *et al.*: localization in two dimensions, a metal–insulator transition in $(2 + \epsilon)$ dimensions.

[11] E. Brézin, S. Hikami and J. Zinn-Justin, Generalized non-linear sigma models with gauge invariance, *Nucl. Phys. B*, **165** (1980) 528.

14

Large n

The physics of the linear and non-linear sigma models is easy to grasp in the large n limit; they do lead to identical critical pictures as expected.

14.1 The linear $O(n)$ model

Let us begin with the linear model; the (Euclidean) action reads:

$$A(\vec{\psi}) = \int \mathrm{d}^d x \left[\frac{1}{2} (\partial_\mu \vec{\psi})^2 + nV(\vec{\psi}^2/n) \right], \tag{14.1}$$

in which V is a simple polynomial such as

$$V(X) = \frac{1}{2} m_0^2 X + \frac{1}{8} g_0 X^2. \tag{14.2}$$

Note that the mass is finite in the large n limit, but the coupling constant of ψ^4 is proportional to $1/n$. In such a theory, the correlation functions of $O(n)$ invariant operators factorize in the large n limit:

$$\langle A[\vec{\psi}^2(x)] B[\vec{\psi}^2(y)] \rangle = \langle A[\vec{\psi}^2(x)] \rangle \times \langle B[\vec{\psi}^2(y)] \rangle + O\left(\frac{1}{n}\right), \tag{14.3}$$

as follows in perturbation theory from counting the number of independent internal indices in Feynman diagrams. The 'master field'

$$\rho(x) = \frac{1}{n} \vec{\psi}^2(x) \tag{14.4}$$

behaves thus as a classical quantity, and therefore it is expected that, if we formulate the theory in terms of this master field, at leading order the theory will be classical. Following Zinn-Justin[1] we implement the definition of ρ through the integration

[1] J. Zinn-Justin, *Quantum Field Theory and Critical Phenomena*, 3rd edn. (Oxford: Oxford University Press, 2002).

$$\int \mathrm{D}\lambda(x)\mathrm{D}\rho(x)\mathrm{e}^{\int \mathrm{d}^d x \frac{1}{2}\lambda(x)(\vec{\psi}^2(x)-n\rho(x))} = 1, \tag{14.5}$$

in which the (properly normalized) integration over $\lambda(x)$ runs over the pure imaginary direction. We have thus

$$A(\vec{\psi},\rho,\lambda) = \int \mathrm{d}^d x \left[\frac{1}{2}(\partial_\mu\vec{\psi})^2 + nV(\rho(x)) + \frac{1}{2}\lambda(x)(\vec{\psi}^2(x) - n\rho(x))\right]. \tag{14.6}$$

One can now trace out the original field $\vec{\psi}$, through a Gaussian integral leading to

$$A(\rho,\lambda) = n\int \mathrm{d}^d x \left[V(\rho(x)) - \frac{1}{2}\lambda(x)\rho(x) + \frac{1}{2}\mathrm{Tr}\log(-\nabla^2+\lambda)\right]. \tag{14.7}$$

The factor n in front of an n-independent functional leads immediately to the saddle-point solution:

$$\begin{aligned} V'(\rho(x)) &= \lambda(x)/2 \\ \rho(x) &= \left\langle x \left| \frac{1}{-\nabla^2+\lambda(x)} \right| x \right\rangle. \end{aligned} \tag{14.8}$$

The symmetry under translation implies uniform solutions $\lambda(x) = \lambda$, and $\rho(x) = \rho$ and thus

$$\left\langle x \left| \frac{1}{-\nabla^2+\lambda} \right| x \right\rangle = \frac{1}{(2\pi)^d}\int \frac{\mathrm{d}^d p}{p^2+\lambda}, \tag{14.9}$$

the integral over p being meant with an ultraviolet cut-off given by a microscopic lattice spacing. This integral is a monotonically decreasing function of λ, which plays the rôle of a renormalized mass squared, and is bounded by its value at $\lambda = 0$

$$\rho_c = \frac{1}{(2\pi)^d}\int \frac{\mathrm{d}^d p}{p^2}. \tag{14.10}$$

At this limiting point the constants in the potential V are constrained by the first equation (14.8) $V'(\rho_c) = 0$, i.e., in the simplest quartic potential,

$$m_{0c}^2 = -\frac{1}{2}g_0\rho_c. \tag{14.11}$$

Unbroken phase, $\rho < \rho_c$

For ρ smaller than ρ_c, one can write

$$\rho - \rho_c = -\frac{\lambda}{(2\pi)^d}\int \frac{\mathrm{d}^d p}{p^2(p^2+\lambda)} \tag{14.12}$$

and the r.h.s. behaves, for small λ, as $\lambda^{(d-2)/2}$. But for ρ close to ρ_c the first equation (14.8) gives

$$\frac{1}{2}(m_0^2 - m_{0c}^2) + \frac{1}{4}g_0(\rho - \rho_c) = \lambda. \tag{14.13}$$

When the correlation length $\xi = \frac{1}{\sqrt{\lambda}}$ diverges, the term $(\rho - \rho_c) \sim \lambda^{(d-2)/2}$ is dominant with respect to λ^1 when $(d-2)/2$ is smaller than one, i.e., when $d < 4$, and the correlation length behaves as $\xi \sim (m_0^2 - m_{0c}^2)^{-1/(d-2)}$. For $d > 4$, the r.h.s. of (14.13) dominates and one recovers the mean field exponent $\xi \sim (m_0^2 - m_{0c}^2)^{-1/2}$.

Broken symmetry

When we have traced out $\vec{\psi}(x)$, we have somewhat hidden the possibility of a broken symmetry. We may exhibit the broken symmetry through several methods: (i) without breaking the symmetry by showing that $\langle \vec{\psi}(x) \cdot \vec{\psi}(y) \rangle$ does not vanish for large $|\vec{x} - \vec{y}|$, its limit being the modulus square of the spontaneous magnetization; (ii) by adding an external field and showing that, if we let the volume go to infinity first, the expectation value of $\vec{\psi}$ does not vanish with the field; (iii) if we single out one component of the field.

Given that we have traced out $\vec{\psi}$ it is easier to use the third method. If we split the field into a longitudinal component $\sigma(x)$ and $(n-1)$ transverse components $\pi_a(x)$, it is clear that in a phase in which the expectation values of the transverse components π_a vanish, the field $\sigma(x)$ is of order $\sqrt{n}$, since $\sigma^2 + \pi_a^2 = n\rho$ and ρ is of order one. Let us thus trace out only the transverse components. We obtain the action

$$S(\sigma, \rho, \lambda) = n \int \mathrm{d}^d x \left[\frac{1}{2N}(\nabla\sigma)^2 + V(\rho(x)) + \frac{1}{2}\lambda(x)(\sigma^2/n - \rho(x)) \right] + \frac{n-1}{2n} \mathrm{Tr} \log(-\nabla^2 + \lambda(x)). \tag{14.14}$$

Since $\sigma(x)$ is a priori of order $\sqrt{n}$ we can again use the saddle-point equations for large n. They are now

$$\begin{aligned} V'(\rho(x)) &= \lambda(x) \\ \rho(x) - \sigma^2(x)/n &= \left\langle x \left| \frac{1}{-\nabla^2 + \lambda(x)} \right| x \right\rangle \\ -\nabla^2\sigma + \lambda(x)\sigma(x) &= 0, \end{aligned} \tag{14.15}$$

in which we have replaced $(n-1)/n$ by one. For a uniform solution one has

$$\lambda\sigma = 0 \tag{14.16}$$

and we see that when λ vanishes there is a phase with non-zero magnetization σ. In that phase the mass (squared) λ of the Goldstone bosons remains zero, and the equations (14.15) become

$$V'(\rho) = 0$$
$$\rho - \sigma^2/n = \rho_c. \tag{14.17}$$

There is now a solution for $\rho > \rho_c$, provided σ is non-zero, i.e., in the phase of broken symmetry. Noting again that

$$\frac{1}{2}(m_0^2 - m_{0c}^2) + \frac{1}{4}g_0(\rho - \rho_c) = 0, \tag{14.18}$$

one has below the critical point (i.e., $m_0^2 < m_{0c}^2$), the spontaneous magnetization σ vanishing like $(m_0^2 - m_{0c}^2)^{1/2}$: the mean field exponent is not modified in the large n limit.

It is interesting though to exhibit the spontaneously broken symmetry without any explicit symmetry breaking. The mathematics are reminiscent of the mechanism of the Bose–Einstein condensation in an ideal gas. If we return to (14.6) we see immediately that

$$\langle \vec{\psi}(x) \cdot \vec{\psi}(y) \rangle = \left\langle x \left| \frac{1}{-\nabla^2 + \lambda} \right| y \right\rangle \tag{14.19}$$

but in view of the massless limit we have to be careful with the volume. Let us imagine that the system is enclosed in a box of volume $\omega = L^d$. With periodic boundary conditions, the free eigenmodes in this box are labelled by momenta $p_a = 2\pi n_a/L,\ a = 1, \ldots, d$ and

$$\left\langle x \left| \frac{1}{-\nabla^2 + \lambda} \right| y \right\rangle = \frac{1}{\omega} \sum_p \frac{e^{i\vec{p}(\vec{x}-\vec{y})}}{p^2 + \lambda}. \tag{14.20}$$

If λ goes to zero we have to isolate the zero-mode $p_a = 0$; in the large-volume limit the sum over the non-zero modes converges to the usual integral and we obtain

$$\left\langle x \left| \frac{1}{-\nabla^2 + \lambda} \right| y \right\rangle = \frac{1}{\lambda\omega} + \int \frac{d^d p}{(2\pi)^d} \frac{e^{i\vec{p}(\vec{x}-\vec{y})}}{p^2 + \lambda}. \tag{14.21}$$

The integral on the right-hand side vanishes exponentially in the massive phase for large $|\vec{x} - \vec{y}|$, and it behaves as $|\vec{x} - \vec{y}|^{-(d-2)}$ in the massless phase. This representation shows that the massless limit and infinite-volume limit do not commute. Therefore, the square of the magnetization is given by

$$M^2 = \lim_{|\vec{x}-\vec{y}| \to \infty} \langle \vec{\psi}(x) \cdot \vec{\psi}(y) \rangle = \frac{1}{\lambda\omega}. \tag{14.22}$$

Similarly, the equation (14.8) reads

$$\rho = \left\langle x \left| \frac{1}{p^2 + \lambda} \right| x \right\rangle = \frac{1}{\lambda \omega} + \int \frac{\mathrm{d}^d p}{(2\pi)^d} \frac{1}{p^2 + \lambda}. \tag{14.23}$$

In the massless phase, in which $\lambda \to 0$ this gives

$$\rho = M^2 + \rho_c, \tag{14.24}$$

which shows that there is indeed a solution with spontaneously broken symmetry if $\rho > \rho_c$.

It is clear that the representation (14.7) allows for a systematic expansion in powers of $1/n$, and several terms, for the expansion in $1/n$ of the critical properties, have been calculated. However, we would like to stress that, together with the expansion in powers of $(4 - d)$, the $1/n$ expansion is not simply a calculational scheme, but mainly a way of defining the critical, i.e., massless theory, in dimensions lower than four. Indeed, it is well-known that a straight perturbation expansion for a massless theory in dimension smaller than four is plagued with infrared divergences. Let us consider, for instance, the 'bubble' diagram

$$B(p) = \int \frac{\mathrm{d}^d k}{k^2 (p - k)^2}. \tag{14.25}$$

Counting dimensions, one sees that, at small p,

$$B(p) \sim p^{-(4-d)}. \tag{14.26}$$

Inserting a string of k such bubbles in a two-point function diagram at momentum q, one obtains for this diagram

$$\int \mathrm{d}^d p \frac{B^k(p)}{(q - p)^2}, \tag{14.27}$$

which diverges in the infrared if

$$k(4 - d) \geq d. \tag{14.28}$$

Therefore, at fixed d this divergence will make the perturbation theory break down at order $k \geq d/(4-d)$. We know of two remedies in the continuum limit described by a field theory:

(a) ϵ expansion: at any order, $\epsilon = (4-d)$ goes to zero first, and therefore one never enters into the dangerous regime of the inequality (14.28).

(b) $1/n$ expansion: the bubble graph also yields a free sum over internal indices, i.e., a sum which is proportional to n. Each additional bubble carries such a factor, together with a coupling constant g_0/n. The large n limit re-sums the geometric series given by the string of bubbles $B(p)$ and replaces it by

$$B(p) \to \frac{B(p)}{1 - g_0 B(p)}, \tag{14.29}$$

which is now finite when p goes to zero. The previous infrared divergence is thus absent in this leading large n limit and the $1/n$ expansion in the critical regime remains well-defined for all dimensions.

14.2 $O(n)$ sigma model

It is interesting to compare the previous analysis with that of the non-linear sigma model in the large n limit. We now normalize the n-component field $\vec{v}(x)$ as $\vec{v}^2 = 1$,

$$A(\vec{v}, \alpha) = \frac{n}{2\lambda} \int \mathrm{d}^d x (\partial_\mu \vec{v})^2 + \frac{n}{2\lambda} \int \mathrm{d}^d x \alpha(x)(\vec{v}^2 - 1), \tag{14.30}$$

in which we have used a Lagrange multiplier $\alpha(x)$ to enforce the constraint, and changed, for convenience λ to λ/n. (The multiplier α should include a factor i; however, it turns out, as above, that its saddle-point value is purely imaginary and we have simply anticipated this fact.) Note that the parameter λ is proportional to the absolute temperature, since the model is simply a large-distance continuum version of a Boltzmann weight for classical spins with short-distance interactions. For large n we will see that a semi-classical approximation becomes exact, once we formulate the problem in terms of degrees of freedom whose number is independent of n.

We may then integrate out the $\vec{v}$ field and obtain:

$$A[\alpha] = -\frac{n}{2\lambda} \int \alpha(x) \mathrm{d}x + \frac{n}{2} \mathrm{Tr} \log [-\nabla^2 + \alpha(x)]. \tag{14.31}$$

Since there is only one degree of freedom $\alpha(x)$, the saddle-point approximation is exact in the large n limit:

$$\frac{2}{n} \frac{\delta A}{\delta \alpha(x)} = -\frac{1}{\lambda} + \langle x | [-\nabla^2 + \alpha(x)]^{-1} | x \rangle = 0. \tag{14.32}$$

Translation invariance allows one to look for a uniform solution ($\alpha(x) = \alpha$):

$$\frac{1}{\lambda} = \int \frac{\mathrm{d}^d p}{(2\pi)^d} \frac{1}{p^2 + \alpha} \tag{14.33}$$

with some ultraviolet cut-off given by the inverse of the lattice spacing.

$d = 2$

In two dimensions,

$$\int \frac{d^2 p}{(2\pi)^2} \frac{1}{p^2 + \alpha} \sim \frac{1}{2\pi} \log \frac{\Lambda}{\sqrt{\alpha}}, \qquad \text{for } \alpha \to 0, \tag{14.34}$$

and thus for any temperature λ there is a non-zero solution. For low T:

$$\alpha \sim e^{-\frac{4\pi}{\lambda}}. \tag{14.35}$$

But $\alpha = m^2 = 1/\xi^2$ (ξ is a correlation length). Therefore, the theory, which began as massless, generated a mass dynamically, as it should, since we know that there is no continuous symmetry breaking in two dimensions. At low T the correlation length ξ is large and behaves as $e^{2\pi/\lambda}$, but it remains finite at any temperature.

$d > 2$

We have a phenomenon mathematically similar to what we found in the linear model. The (cut-off) integral

$$\int \frac{d^d p}{(2\pi)^d} \frac{1}{p^2 + \alpha}$$

reaches its maximum at $\alpha = 0$. Calling this maximum

$$\frac{1}{\lambda_c} = \int \frac{d^d p}{(2\pi)^d} \frac{1}{p^2}, \tag{14.36}$$

we see that the saddle-point equation has a solution only for $\frac{1}{\lambda} < \frac{1}{\lambda_c}$, i.e., for $T > T_c$. For those temperatures, we can find the behaviour of $\alpha = 1/\xi^2$ when T approaches T_c. Indeed,

$$\frac{1}{\lambda_c} - \frac{1}{\lambda} = \alpha \int \frac{d^d p}{(2\pi)^d} \frac{1}{p^2(p^2 + \alpha)}. \tag{14.37}$$

The right-hand side of this equation vanishes with α as $\alpha^{1+d/2-2} = \alpha^{d/2-1} = \xi^{-(d-2)}$. The correlation length is thus diverging at λ_c as

$$\xi \sim (\lambda - \lambda_c)^{-\nu}, \qquad \nu = \frac{1}{d-2}, \tag{14.38}$$

as found earlier with the linear model. What happens if the temperature is such that $\lambda < \lambda_c$, since there is no solution to the saddle-point equation (14.33)? Again, we have to return to the derivation and either allow for a broken symmetry or show that the two-point function $\langle \vec{v}(x)\vec{v}(y) \rangle$ does not vanish at large separations at temperatures lower than T_c. If we choose to split again $\vec{v}$ into a longitudinal field

$\sigma(x)$ and transverse components $\pi_a(x)$, we trace out the $(n-1)$ π-modes. We obtain then the action

$$A(\sigma,\alpha)=\frac{n}{2\lambda}\int \mathrm{d}^d x\left[(\partial_\mu\sigma)^2+\alpha(\sigma^2-1)\right] \\ +\frac{1}{2}(n-1)\operatorname{Tr}\log(-\nabla^2+\alpha). \tag{14.39}$$

Again, for large n the solution is given by a saddle-point, i.e., neglecting the difference between n and $(n-1)$, by

$$-\nabla^2\sigma+\sigma\alpha=0 \\ \lambda\left\langle x\left|\frac{1}{-\nabla^2+\alpha}\right|x\right\rangle=1-\sigma^2, \tag{14.40}$$

which, with translational invariance, gives

$$\sigma\alpha=0 \\ \lambda\int\frac{\mathrm{d}^d p}{(2\pi)^d}\frac{1}{p^2+\alpha}=1-\sigma^2. \tag{14.41}$$

The broken symmetry phase is again apparent with $\alpha=0$, i.e., massless Goldstone bosons and the second equation reading

$$\frac{\lambda}{\lambda_c}=1-\sigma^2. \tag{14.42}$$

We discover again that the spontaneous magnetization σ vanishes at the critical point as $(T_c-T)^{1/2}$. This physical picture persists for finite n.

It is interesting to compare the renormalization group strategies of the linear and non-linear model. The linear theory involves one relevant variable, the mass squared, and one irrelevant variable in dimensions less than four, the φ^4 coupling constant, which is driven to its fixed point in the large-distance limit. The equivalent non-linear model contains only the relevant temperature variable, which has to be fixed to its critical value to obtain a large-distance scale-invariant theory. Therefore, the non-linear model fixed point for its temperature coupling constant is necessarily an ultraviolet fixed point. Let me stress again that, since the relevant temperature variable flows away from the fixed point in the large-distance limit, it is an infrared repulsive fixed point, which is identical to an ultraviolet stable fixed point when there is only one variable. In two dimensions, the model has to flow away from the zero-temperature fixed point in the infrared: the non-linear sigma model is thus certainly asymptotically free (in the ultraviolet), as shown in the previous chapter from explicit calculations.

Finally, we have seen that the linear and non-linear models are equivalent descriptions of the same physics in the large n limit. In fact, this persists to all orders in $1/n$. Furthermore, there is complete consistency between the $(4-d)$, $1/n$ and $(d-2)$ expansions. For instance, the critical exponent ν, which governs the divergence, of the correlation length, is given by the three expansions

$$\begin{aligned}\nu &= \frac{1}{2}+\frac{(n+2)(4-d)}{4(n+8)}+O(4-d)^2\\ &= \frac{1}{d-2}-\frac{2}{n}\frac{\Gamma(d)}{d\,\Gamma^3(d/2)\Gamma(2-d/2)}+O(1/n^2)\\ &= \frac{1}{d-2}-\frac{1}{n-2}+O(d-2).\end{aligned} \tag{14.43}$$

Several more terms are known in those three expansions, but I simply quote the results to stress their compatibility. For instance, if one takes the large n result (middle line) and expands it for d near four, one recovers the first line with $(n+2)/(n+8)$ replaced by $(1-6/n)$; the same is true if one considers large n with d near two, provided $1/(n-2)$ is replaced by $1/n$ in the $(d-2)$ expansion.

In conclusion, one sees that, if computing accurately, the critical properties of a given system in three dimensions remain a problem requiring delicate non-perturbative expansions; the coherence of the picture in the d, $1/n$ plane gives confidence that no essential physical element is missing in our understanding of the problem. One can say that it is well-understood in the same way that quantum electrodynamics are understood.

Index

Printed by Integrated Books International,
United States of America